75

John Dalton

Space Travellers
The Bringers of Life

SPACE TRAVELLERS

The Bringers of Life

FRED HOYLE
CHANDRA WICKRAMASINGHE

Edited by Barbara Hoyle

University College Cardiff Press

First published 1981 in Great Britain by
University College Cardiff Press,
P.O. Box 78, Cardiff, CF1 1XL, United Kingdom.

British Library Cataloguing in Publication Data
Hoyle, Fred
1. Cosmology
I. Title
II. Wickramasinghe, Chandra
III. Hoyle, Barbara
523.1'01 QB981

ISBN 0 906449 27 8

Printed in Wales by CSP Printing of Cardiff.

In memory of
Svante Arrhenius

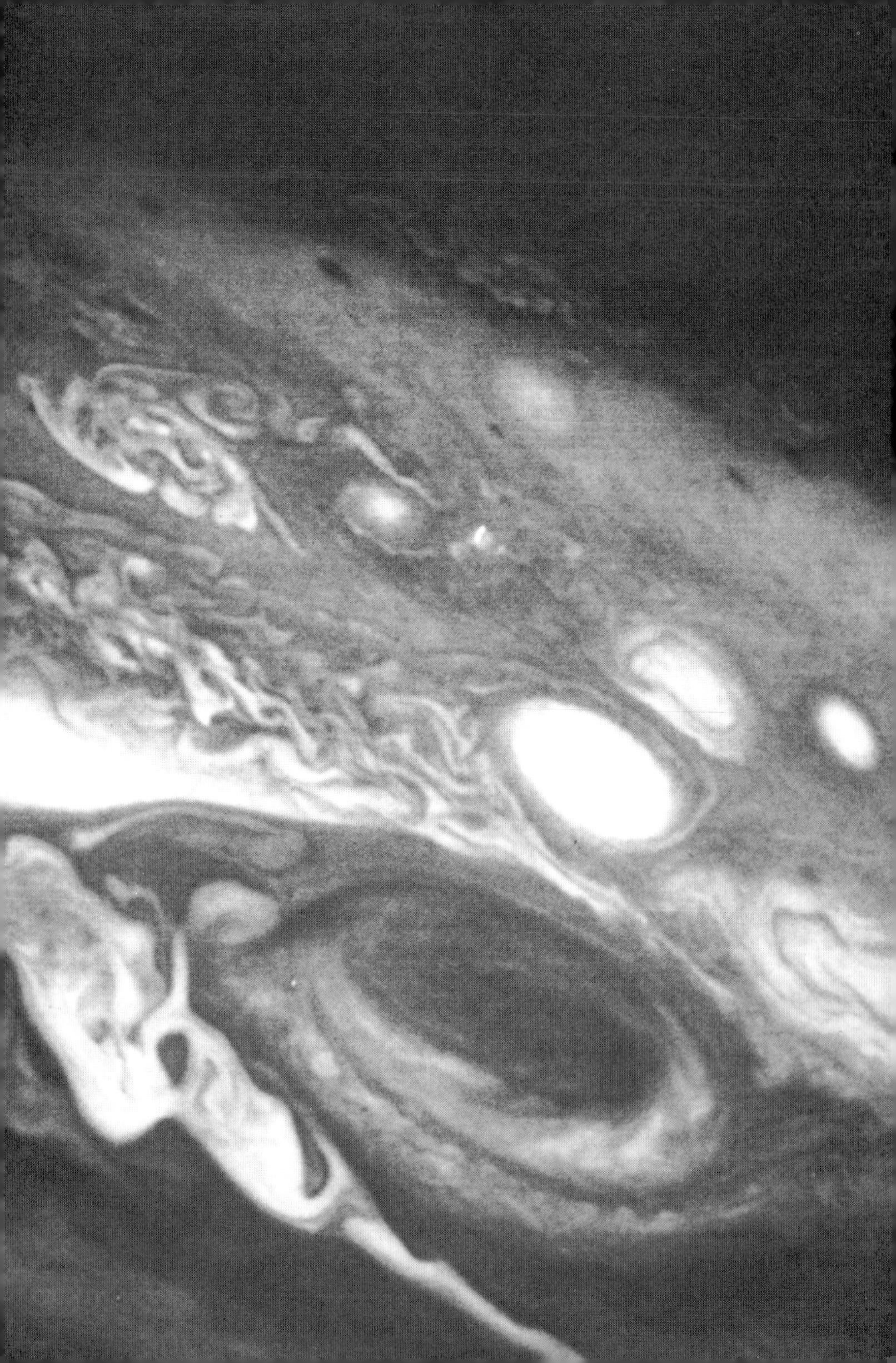

CONTENTS

PHOTOGRAPHS

Chapter page illustrations and acknowledgements

1

WRIT IN THE STARS

When sightings of unidentified flying objects were claimed in the years following the second world war it seemed that genuine occurrences, like the reflection of sunlight from an escaped weather balloon, would soon be separated from the rest, the rest being nothing but a motley collection of inaccurate observation, hysterical reports, and deliberately planned fakes.

This did not happen, however. In 1950 people wanted to believe in UFOs, but they didn't. Nowadays, people want to believe in UFOs, and they do. There is a widespread feeling that the time is ripe for something of the kind to be proved true, and in a certain sense we shall demonstrate in this book that it is indeed true.

However, the Earth is not being invaded by intelligent beings in spacecraft. Such a concept is but a crude perception of the real situation; it is to 'see through a glass darkly', just as the poet who is struck by the beauty of a rose or the majesty of the starry sky perceives only glimpses of far deeper wonders.

A rose carries within it a heritage that stretches across aeons of time, a heritage which connects its fleeting present existence to the most remote past, and to distant far-flung parts of the Milky Way — our galaxy. So it is for every speck of living matter on the Earth, including of course ourselves. In the starry vault above us is writ, not merely the story of life on our planet, but of life throughout the universe.

The oxygen, carbon, nitrogen, phosphorus, sulphur, and the metals (to name several of the more important kinds of atom of which living material is composed) had their origin in the stars. This much is well-known. In this book we shall go further by showing that the exceedingly complex arrangements of these atoms in living material also had their origin on a cosmic scale. The Swedish chemist, Svante Arrhenius, whose work we shall review in the next chapter, was the first to develop this idea — an ancient idea — in a serious manner. We shall find that recent developments in astronomy and biology make the case for life as a galaxy-wide phenomenon quite overwhelming.

Let us here state the questions for which we seek the answers:

'How frequently might we expect life to turn up in the universe if we could somehow make such a survey?'

'How uniform is the basic chemical fabric of life?'

We shall find that answers to these questions, which have long been thought unanswerable, became almost self-evident as soon as we put together the known facts in an appropriate way.

2
SVANTE ARRHENIUS

Svante August Arrhenius was born at Vik, in Sweden, on 19 February 1859, the son of a surveyor and land agent. He is chiefly remembered for his work on electrolytes.

It is a matter of everyday knowledge that, if the liquid from a charged car battery is emptied and the battery is refilled with distilled water only, little in the way of an electric current can be obtained from it. The reason is that pure water is a poor conductor of electricity. If, however, some acid — usually sulphuric acid — is added to the distilled water, the battery immediately delivers its current as usual. The acid which thus plays a crucial role in transmitting electricity through water is known as an electrolyte. Common salt, which is made up of molecules each consisting of an atom of sodium and an atom of chlorine, is also an electrolyte.

As a student at Uppsala University Arrhenius suggested in his doctorate thesis that electrolytes dissociate on solution in water. This means that the molecules come apart, as for instance the sodium ion separates from the chlorine ion in the case of molecules of common salt. The separation is of a peculiar kind, however, with the chlorine taking an electron away from the sodium to form charged ions. Arrhenius did not know about electrons at that time (1883), but he suggested that when the molecules of an electrolyte dissociate into two pieces, the pieces are equally and oppositely charged, one positive, the other negative, and that it is the charging of the pieces which permits electricity to flow through the solution.

Correct as his theory was, the young student soon got into controversy. Disbelieved by most of the senior chemists in Sweden, his thesis received the lowest passing grade that the University of Uppsala could bestow. But Arrhenius seems to have been a stubborn fellow, not easily put down by this rebuff. He circulated copies of his work to leading scientists throughout Europe, and as the years passed the theory gradually became more and more accepted. Eventually it was judged respectable enough for Arrhenius to be elected a member of the Swedish Academy of

Sciences. Two years later, in 1903, he received the Nobel Prize for chemistry.

Having now come successfully through one scientific dogfight, Arrhenius almost immediately started another. This second controversy, which began with the publication of a remarkable book, *Worlds in the Making,*[1] he was not destined in his lifetime to win. We shall show in this book, however, that the crucial ideas of *Worlds in the Making*, those of its last chapter, were well-argued and, we think, entirely correct in principle, if not in precise detail.

Biology has paid a heavier price than astronomy for ignoring Arrhenius. It has been argued that, because our only direct experience of life is here on the Earth, life must have originated on the Earth. However, although Gaelic is spoken as a native language only in certain parts of the West of Scotland and the West of Ireland, it did not originate in these regions. Gaelic is a language more than 3000 years old, and it came to the British Isles from Europe. The presumption that life originated on the Earth is quite without logical foundation. Nevertheless, scientists have complacently accepted this myopic point of view for many decades. The result has inevitably been disastrous. This Earth-centred view has led to many contradictions and confusions in present-day biological thought.

Ironically, the ideas developed in *Worlds in the Making* come from the strict acceptance of one of the most important doctrines of nineteenth century biology. The doctrine stated in a simple form is that life can only be derived from life. It was Arrhenius who accepted this doctrine, and it has been his Earth-centred opponents who have persistently broken it. Louis Pasteur enunciated the doctrine when he told the French Academy of Sciences that the concept of spontaneous generation would not survive the 'mortal blow' which he had delivered to it by certain carefully designed experiments.

[1] Arrhenius, S., *Worlds in the Making* (New York and London: Harper & Row, 1908).

The concept of spontaneous generation, according to which life originates in mixtures of simple materials — earth, air, water, had persisted from the time of Aristotle (384–322 BC) up to the middle of the nineteenth century. Of course, nobody had seriously thought that big animals originated in this way. It was accepted that a calf can be obtained only from a cow, although in Shakespeare's *Antony and Cleopatra* Lepidus tells Mark Antony:

> 'Your serpent of Egypt is born in the mud, by the action of the Sun, and so is your crocodile.'

Small creatures, such as fire-flies, worms, and maggots, were widely believed, however, to originate spontaneously out of simple inorganic materials.

Many had been the experiments that claimed to demonstrate the spontaneous emergence of life, but always when such experiments were repeated with better precautions the claimed results were shown to be incorrect. Already in 1668 the Italian physician Francesco Redi had shown that maggots would not form in meat if the precaution of keeping flies away from it were taken. It became clearer and clearer that in all cases where the spontaneous emergence of life was supposed to happen some form of living precursor, usually a tiny egg, was always present. Even the smallest creatures were like the calf and cow. Each generation of every animal was preceded by a generation of the same animal. This was the doctrine which Louis Pasteur enunciated more than a century ago to the French Academy of Sciences.

Yet by a remarkable piece of mental gymnastics biologists were still happy to believe that life started on the Earth through spontaneous processes. Each generation was considered to be preceded by a similar generation, but only back so far in time. Somewhere along the chain was a beginning, and the beginning was a spontaneous origin. Thus spontaneous generation was both accepted and rejected.

The contradictory nature of this reasoning was softened a little by the Darwin–Wallace theory of natural selection. According to this theory, complex animals evolve slowly over the generations from simpler ones. It was argued that in the beginning the first animal could therefore have been exceedingly simple, and the spontaneous generation of an initially very simple creature did not seem so much of a problem as it would for the more complex creatures that we have around us today. We shall discuss in another book[1] whether this argument of primitive forms growing into complex forms was really correct or not. Suffice it to say here that it is an argument which has satisfied most scientists right up to the present-day.

Most, but not all. There were a few scientists already in the nineteenth century who felt the situation to be contradictory. If spontaneous origin could not happen, as Louis Pasteur had claimed to the French Academy, then it could not happen. Every generation of every living creature had to be derived from a preceding generation, going backward in time to a stage before even the Earth itself existed. Hence it followed that life must have come to the Earth from outside, from a previous existence somewhere else in the universe.

This argument is quite powerful enough for it to have received respectful consideration by biologists and astronomers. There were of course subsidiary questions to be answered. How could life move from one planet to another? How could it cross the immensity of space from one star system to another?

The first attempts to answer these questions were crude. H.E. Richter, a German physician, argued that not all the chunks of iron and stone which enter the Earth's atmosphere from time to time (meteorites) plunge to the ground. Some meteorites must encounter the atmosphere at glancing angles, penetrating only a little way into the air, and then going out again, away from the Earth. Such meteorites might pick up living cells from the atmosphere, Richter argued, carrying them out into space to whatever destination the meteorites might eventually reach.

[1] *Evolution from Space* (London: J.M. Dent, 1981).

On a more ambitious scale, William Thompson (later Lord Kelvin) remarked in his presidential address to the 1881 meeting of the British Association:

> 'When two great masses come into collision in space, it is certain that a large part of each is melted, but it seems also quite certain that in many cases a large quantity of debris must be shot forth in all directions, much of which may have experienced no greater violence than individual pieces of rock experience in a landslip or in blasting by gunpowder. Should the time when this earth comes into collision with another body, comparable in dimensions to itself, be when it is still clothed as at present with vegetation, many great and small fragments carrying seeds of living plants and animals would undoubtedly be scattered through space. Hence, and because we all confidently believe that there are at present, and have been from time immemorial, many worlds of life besides our own, we must regard it as probable in the highest degree that there are countless seed-bearing meteoric stones moving about through space. If at the present instant no life existed upon the earth, one such stone falling upon it might, by what we blindly call *natural* causes, lead to its becoming covered with vegetation.'

It is the remarkable merit of the final chapter of *Worlds in the Making* that it lifted these quite undeveloped ideas of Richter and Kelvin to the level of a serious scientific theory without immediate technical weaknesses.

The key to the thinking of Arrhenius lay in the phenomenon of radiation pressure. When light or heat is absorbed or reflected by any body, a pressure is exerted on the body. For large objects, radiation pressure is usually very small, but for small particles the pressure of radiation may be larger than any other force acting on them.

Arrhenius would seem to have had a poor opinion of most astronomers:

> 'The astronomers followed faithfully in the footsteps of their inimitable master, Newton, and they brushed aside every

> phenomenon which would not fit into his system. An exception was made by the famous Euler, who, in 1746, expressed the opinion that the waves of light exerted a pressure upon the body upon which they fell. This opinion could not prevail against the criticisms with which ... others assailed it. That Euler was right, however, was proved by Maxwell's great theoretical treatise on the nature of electricity (1873).'

After a short digression, Arrhenius comes next to the essence of his argument:

> 'In spite of Maxwell's great authority, astronomers quite overlooked this important law of his [the law which demonstrated the pressure of radiation]. Lebedeff, indeed, tried in 1892 to apply it [radiation pressure] to the tails of comets, which he regarded as gaseous; but the law is not applicable in this case. As late as the year 1900, shortly before Lebedeff was able to publish his experimental verification of this law, I attempted to prove its vast importance for the explanation of several celestial phenomena. The magnitude of the radiation pressure of the solar atmosphere must be equivalent to 2.75 milligrammes if the rays strike vertically against a black body one square centimetre in area. I also calculated the size of a spherule of the same specific gravity as water, such that the radiation pressure to which it would be exposed in the vicinity of the sun would balance the attraction by the sun. It resulted that equilibrium would be established if the diameter of the sphere were 0.0015 millimetres (mm). A correction supplied by Schwarzschild showed that the calculation was only valid when the sphere completely reflects all the rays which fall upon it. If the diameter of the spherule be still smaller, the radiation pressure will prevail over the attraction, and such a sphere would be repelled by the sun. Owing to the refraction of light, this will, according to Schwarzschild, further necessitate that the circumference of the spherule should be greater than 0.3 times the wavelength of the incident rays. When the sphere becomes still smaller, gravitation will predominate once more. But spherules whose sizes are intermediate between these two limits will be repelled. It results, therefore, that molecules, which have far smaller dimensions than

those mentioned, will not be repelled by the radiation pressure, and that therefore Maxwell's law does not hold for gases. When the circumference of the spherule becomes exactly equal to the wavelength of the radiation, the radiation pressure will act at its maximum, and it will then surpass gravity not less than nineteen times. These calculations apply to all spheres, totally reflecting the light, of a specific gravity like water, and to a radiation and attraction corresponding to that of the sun. Since the sun's light is not homogeneous (of a single wavelength), the maximum effect will somewhat be diminished, and it is nearly equal to ten times gravity for spheres of a diameter of about 0.00016 mm.'

This passage, occurring in the fourth chapter of *Worlds in the Making*, is preparatory to the final chapter in which Arrhenius emphasizes that many living cells have diameters in the range of 0.00016 mm to 0.0015 mm. Such cells, if they could become free of the Earth in some way, would therefore be expelled altogether from the solar system. Arrhenius calculates that they could be expelled at outward speeds of as much as 100 km s^{-1}, and that they could make the journey to the nearest sunlike system (Alpha Centauri) in about 9000 years.

He next worries about two further problems. Once outside the solar system, could living cells retain their germinating power for as long as 9000 years? Could they escape being killed by ultraviolet light while still comparatively close to the Sun at the beginning of their journey? To the first question he answers confidently in the affirmative. His argument is partly experimental and partly theoretical:

'When the spores have passed the orbit of Neptune, their temperature will have sunk to −220° [Centigrade], and farther out it will sink still lower. In recent years experiments have been made in the Jenner Institute, in London, with spores of bacteria which were kept for twenty hours at a temperature of −252° in liquid hydrogen. Their germinating power was not destroyed thereby.

'Professor Macfadyen has, indeed, gone still further. He has demonstrated that micro-organisms may be kept in liquid air (at

−200°) for six months without being deprived of their germinating power. According to what I was told on the occasion of my last visit to London, further experiments, continued for still longer periods, have only confirmed this observation.

'There is nothing improbable in the idea that the germinating power should be preserved at lower temperatures for longer periods than at our ordinary temperatures. The loss of germinating power is no doubt due to some chemical process, and all chemical processes proceed at slower rates at lower temperatures than they do at higher. The vital functions are intensified in the ratio 1:2.5 when the temperature is raised by 10 °C (18 °F). By the time the spores reached the orbit of Neptune and their temperature had been lowered to −220°, their vital energy would, according to this ratio, react with one thousand millions less intensity than at 10°. The germinating power of the spores would hence, at −220°, during the period of three million years, not be diminished to any greater degree than during one day at 10°. It is, therefore, not at all unreasonable to assert that the intense cold of space will act like a most effective preservative upon the seeds, and that they will in consequence be able to endure much longer journeys than we could assume if we judged from their behaviour at ordinary temperatures.'

As regards the second of the above two questions, concerning the injurious effect of solar ultraviolet light before reaching the orbit of Neptune, Arrhenius argues as follows:

'On the road from the earth the germs would for about a month be exposed to the powerful light of the sun, and it has been demonstrated that the most highly refrangible rays of the sun [i.e. ultraviolet] can kill bacteria and their spores in relatively short periods. As a rule, however, these experiments have been conducted so that ... (the spores were in a condition to germinate). That, however, does not at all conform to the conditions prevailing in planetary space. For Roux has shown that anthrax spores, which are readily killed by light when the air has access, remain alive when the air is excluded ... All the botanists that I have been able to consult are of the opinion that we can by no means assert with certainty that spores would be killed by the light rays in wandering through infinite space.'

On approaching another star, radiation pressure from that star could check the speed of a travelling cell. In the event that the speed were checked at just the distance from the star at which a planet happened to be, a gentle encounter with the atmosphere of the planet could ensue, with the cell surviving addition without becoming unduly heated. Conscious that this step of his argument involved a coincidence, Arrhenius says further that a fraction of the cells, after having had their speed checked by radiation pressure, might become attached to inorganic particles appreciably larger than 0.0015 mm, and from there on would be freed from the control of radiation pressure. The inorganic particles with their microbial living attachments could then find their way to the planets of the alien star system.

To complete the picture it is necessary to return to the escape of living cells from the Earth. We have seen that first Richter, and then Lord Kelvin, had made crude suggestions of how this might happen. Arrhenius dismisses these suggestions and offers instead a far more refined theory, according to which living cells are carried high in our atmosphere by rising air currents, and they are then expelled from the top of the atmosphere by electrical forces:

'An air current of a velocity of 2 metres per second would take them [the spores] to a height where the air pressure is only 0.001 mm — i.e. to a height of about 100 km. But the air currents can never push the spores outside of our atmosphere.

'In order to raise them to still higher levels we must have recourse to other forces, and we know that electrical forces can help us out of almost any difficulty [!] At heights of 100 km the phenomena of the radiating aurora take place. We believe that the aurorae are produced by the discharge of large quantities of negatively charged dust coming from the sun. If, therefore, the spore in question should take up negative electricity from the solar dust during an electric discharge, it may be driven out into the sea of ether by the repulsive charges of the other particles.

'We suppose, now, that the electrical charges — like matter — cannot be subdivided without limit. We must finally come to a minimum charge, and this charge has been calculated at about 3.5×10^{-9} electrostatic unit.

'We can, without difficulty, calculate the intensity of an electric field capable of urging the charged spore of diameter 0.00016 mm upward against the force of gravity. The required field strength is only 200 volts per metre. Such fields are often observed on the surface of the earth with a clear sky, and they are, indeed, almost normal. The electric field of a region in which an auroral display takes place is probably much more intense, and would, without doubt, be of sufficient intensity to urge the small electrically charged spores which convection currents had carried up to these strata, farther out into space against the force of gravity.'

We have purposely not commented at this point on the strengths and weaknesses in the chain of the Arrhenius argument. From seventy years on, we shall examine the links in the next two chapters. Here we emphasize the astonishing ingenuity with which the argument was put together. With a certain crucial modification to be discussed in later chapters, we believe it to be correct.

3

SPONTANEOUS GENERATION AND THE ORGANIC SOUP

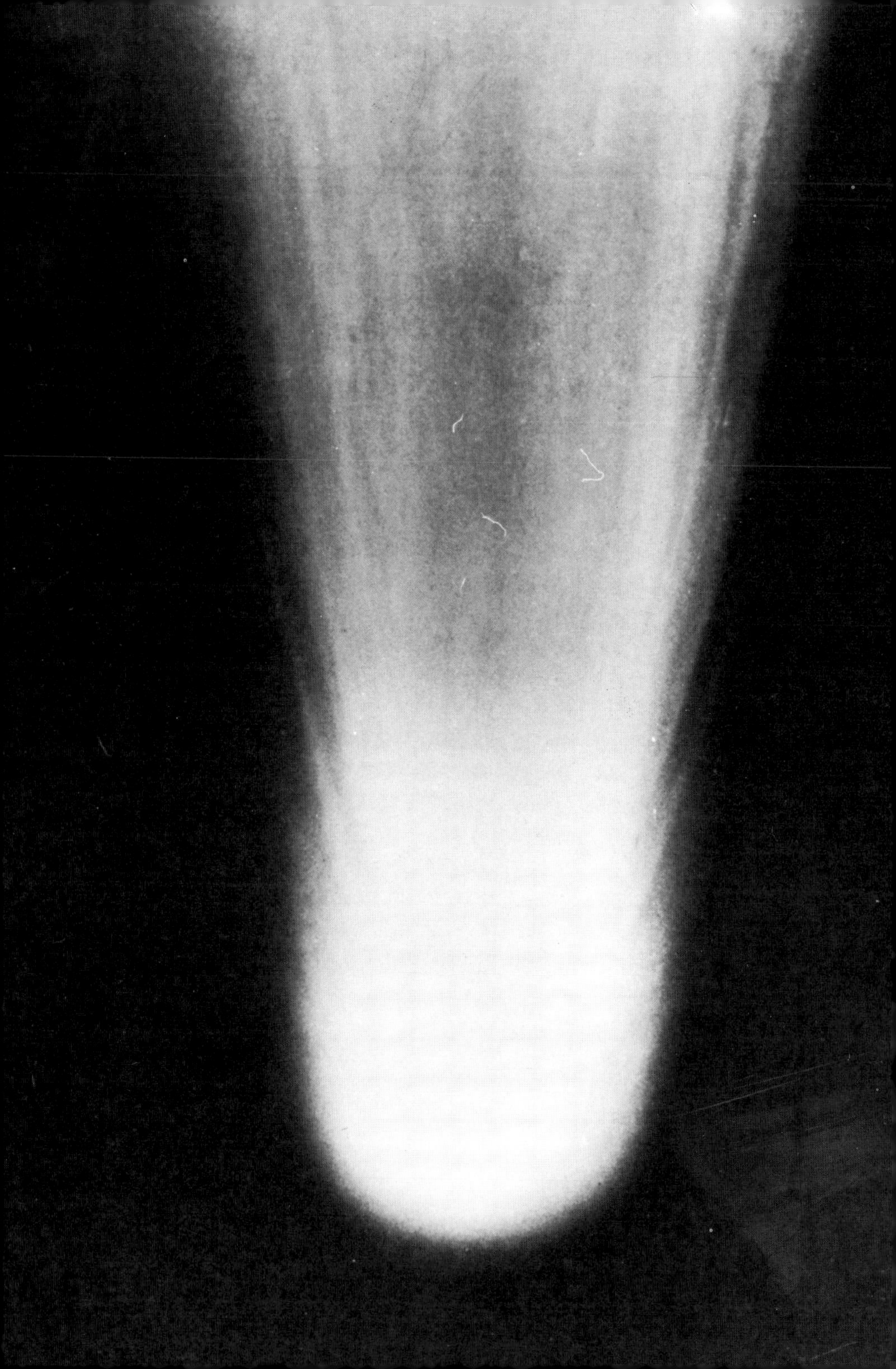

Viewed with the perspective of our present knowledge, we can see that the argument of the previous chapter is stronger at some places than even Arrhenius could have expected. It is weaker at other places, explicitly at three, of which we shall consider one in this chapter and two in the next.

Transferring life from a restricted site on the Earth alone to the universe at large does not avoid the need for a beginning, because present cosmological theory requires all observed aspects of the universe to have had a beginning within the past 10 to 15 billion years.[1] It can be seen from Fig. 3.1 that life on the Earth is at least

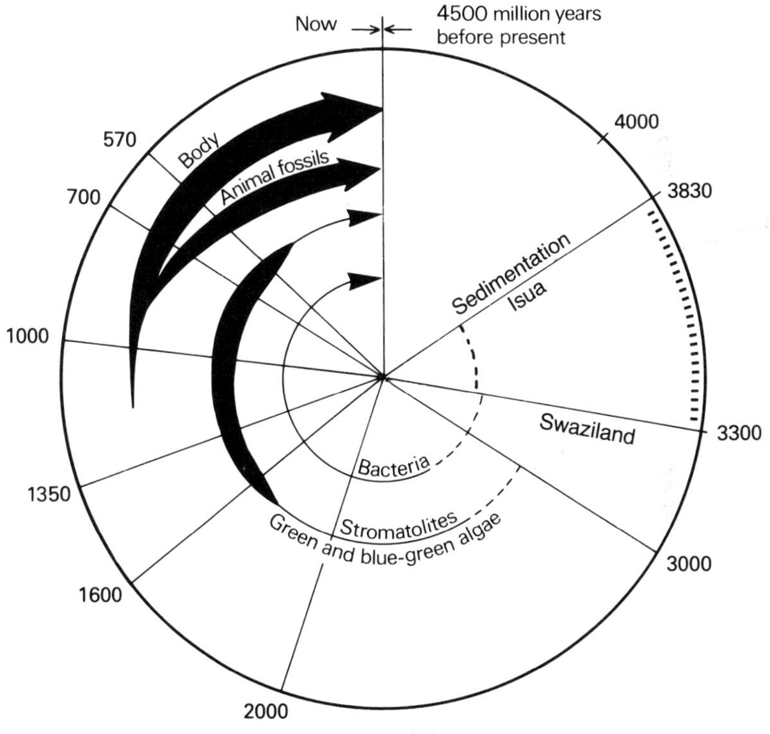

Figure 3.1
Life forms in the geological record.

[1] 1 billion = 1000 million.

3.5 billion years old, so that even the largest scale features of the universe are not more than about four times older than terrestrial life. The statement by Lord Kelvin,

> 'Hence, and because we all confidently believe that there are at present, *and have been from time immemorial* [out italics], many worlds of life beside our own ...'

is therefore open to doubt at its most crucial point. According to present astronomical theory, there may not have been worlds like our own from time immemorial.

This of course assumes a big-bang cosmology. If, however, the broad form of the steady-state cosmology discussed by one of the present authors (*Steady-State Cosmology Revisited*, University College Cardiff Press, 1980) is correct, then the time-scale, while still restricted in some degree, could be much longer than 15 billion years.

We next cite an instance where the theory gains a fundamental point of technical biology. We remarked in the previous chapter that biologists accept Louis Pasteur's denunciation of a spontaneous generation of present-day life forms on the one hand, and are yet happy to believe that life itself started on the Earth through spontaneous processes. We remarked that the contradictory nature of this reasoning has been softened in conventional biological lore by the Darwin–Wallace theory of natural selection. According to this theory, complex animals evolve slowly over long periods of time from simpler ones. In the beginning, the first creature could be very simple, it is argued, so that a spontaneous generation of the first creature is not supposed to be as much of a problem as it would be for the more complex creatures that we have around us today. This argument is badly wrong, however.

The biochemistry of bacteria and viruses is similar to the biochemistry of the individual cells of higher animals such as ourselves, otherwise we would not be subject to attack by bacteria and viruses. It is true that the cells of higher animals are bigger than bacteria, but the more complex structures of these so-called eukaryotic cells look as though they have been achieved by adding smaller pieces together, the smaller pieces being quite likely bacteria and viruses.

In a fundamental chemical sense the oldest life forms on the Earth, the bacteria or yeast-like cells shown in Fig. 3.1, are not simpler than the cells of man. Indeed some of the most important functions of the cells of higher animals, involving a class of proteins known as enzymes, are better studied in the laboratory with the aid of enzymes taken from microorganisms than they are with their native enzymes — our cells can be made to work in various critical respects using bacterial enzymes instead of our own.

It is not hard to find biochemists who will assert that viruses, so far from being chemically more primitive than the cells of man, are actually more subtle. In fact, all life forms operate in the same basic way. What the natural selection theory of Darwin and Wallace has done is to permute and combine with increasing complexity the same basic chemical units. There is no evidence that natural selection, operating on the Earth, has done much, if anything, to increase the complexity of the underlying units.

Both sides of the spontaneous generation argument have therefore been weakened in their positions by modern knowledge. For the followers of Kelvin and Arrhenius there is no escape from spontaneous generation occurring somewhere in the universe. For those who think that spontaneous generation occurred on the Earth there is no escape from facing the problem in essentially its full complexity.

There is not a single fact to support the view that life started on the Earth, whereas there are quite a number of facts to suggest that it did not. We met one such fact in the previous chapter, and we shall discuss others in the next chapter. Arrhenius himself emphasized that seeds germinate after exposure to exceedingly low temperatures. This property was essential to his theory, and he seems to have caused the experiments he describes to be made in order to test the theory. If life originated on the Earth, on the other hand, there is no requirement for survival down to the temperature of liquid hydrogen (−252 °C), since no such very low temperatures have ever been experienced on the Earth. The ability of the basic cellular building blocks of life to withstand extreme cold is fortuitous in the Earth-bound theory, and this is a weak position.

It is remarkable that over the past half-century the scientific world has, almost without exception, believed in a theory for which there is not a single supporting fact. The arguments have been theoretical, not factual. If such arguments had been cogent, the situation might perhaps be excusable. But all the cogency of theoretical argument is *against* the Earth-bound theory, not in its favour. We review the situation briefly.

Life is based on molecules containing principally the atoms of hydrogen, carbon, nitrogen and oxygen, with carbon playing a dominant role in determining the chemical properties of the molecules. Molecules of carbon and nitrogen in association with hydrogen, as in methane, ammonia and hydrogen cyanide, are said to be 'reduced'. Molecules in association with oxygen, as in carbon dioxide, and nitric oxide, are said to be 'oxidized'. The important molecules of life, such as sugar, are neither reduced nor oxidized. They are intermediate, a kind of half-way house. It is life itself which maintains this intermediate situation. Without life, all carbon on the Earth would quickly become oxidized, as when wood or grass is burnt. Nitrogen, if initially in reduced molecules like ammonia, would end up as nitrogen gas, such as in fact constitutes four-fifths of the Earth's atmosphere.

It is not possible to prevent the inexorable trend of carbon towards oxidation merely by the device of denying the presence of oxygen. Even if in the early days of the Earth there was no free oxygen gas, there had to be oxygen combined with hydrogen as water. Sunlight dissociates water into hydrogen and oxygen. Hydrogen, being a very light gas, escapes quickly from the Earth's gravity, going out into space and so leaving behind oxygen gas, which would soon oxidize the carbon. The trend of carbon towards oxidation is therefore inevitable.

The problem for the spontaneous generation of life on the Earth is now severe. Even if the Earth started with reduced molecules of carbon, there would be little prospect of producing life during their brief transition to oxidized forms. Attempts have been made to delay the burning up of carbon by appealing to solar ultraviolet light to break up molecules which become too oxidized, as for instance breaking up carbon dioxide. But such an appeal would be more damaging than helpful, because ultraviolet light would also prevent the basic biochemical structure of life from ever being generated.

Time is also something of a difficulty — a difficulty that has been greatly exacerbated by the discovery of life in the very oldest sedimentary rocks on the Earth. For the present we note that the oldest life on the Earth is now dated about 3.83 billion years ago, whereas the Earth itself has an age of about 4.5 billion years. There is a similar time gap between the origin of the Moon and the ages of the oldest lunar rock formations, suggesting that, until about 3.9 billion years ago, the surfaces of both the Earth and Moon were highly disturbed and unstable, and most probably inhospitable to life. Thus the most recent data indicates that life appeared on the Earth at the very first moment that conditions became favourable for its survival. There is scarcely any time left now for spontaneous generation on the Earth. The so-called 'primordial soup' would seem to have been pretty well squeezed out of any possible existence it might have had in the geological record.

The biochemistry of life, with its intricate relation of proteins to genetic structure, with its subtle use of sunlight to prevent excessive oxidation, appears to us a far too complex phenomenon to be entrusted to the inhospitable habitat which the Earth must have presented over the first billion years of its history, a habitat consisting either of entirely bare ground or of a vast ocean in which organic materials would have become hopelessly diluted into a thin and useless 'primordial soup'. What the theory of spontaneous generation on the Earth asks us to believe is that in this unpromising environment, with the inexorable trend towards the oxidation, a biochemical miracle occurred. Everything of basic biochemical importance occurred then, after which the biochemical inventiveness of life has been essentially zero. This is surely not a picture on which it would be wise to stake much of value.

Rather does it seem to us as if the physical resources of the whole of our galaxy are needed to produce the basic structure of the life we find on Earth today. Only if the biochemical discoveries of thousands of millions of star systems are pooled together in an on-going evolutionary process could the system we find have developed. The magnitude of the biochemical problem is enormous, and it needs enormous resources for its solution.

4

SURVIVAL KIT FOR SPACE TRAVEL

Chemical reactions proceed far more slowly at very low temperatures than at ordinary temperatures, just as Arrhenius pointed out. The slowing of chemical reactions at low temperatures is a consequence of the physical laws, however, not of biological evolution. Superficially it might therefore seem as if the observed ability of seeds immersed in liquid hydrogen to retain their germinating power (see Chapter 2) was not really a biological issue at all. If this were so, life systems that had evolved in a purely terrestrial environment would necessarily have this observed property, and an argument given in the previous chapter in favour of the Arrhenius theory would collapse.

But a cow exposed to very low temperature does *not* retain viability, even though chemical reactions slow just as much in the cow as in a seed. Hence the slowing of chemical reactions is not itself a guarantee (as Arrhenius seems to have thought) that living systems shall have the ability to withstand great cold. Other factors are also involved, as for example the ability to withstand a change from liquid to solid, and *vice versa*, to withstand thawing. Modern experience with freeze-drying techniques shows that the survival even of bacteria, let alone a cow, is by no means guaranteed. The chemical composition of the atmosphere in which the bacteria are suspended, the humidity, and the speed of freezing and thawing are all contributory factors. The relevant issue therefore is whether the combination of all the factors to be expected in space are favourable or unfavourable to survival.

Zero humidity, absence of oxygen, and low temperatures are the conditions of space. This combination has indeed been found by experiment to be highly favourable to survival. The argument of the previous chapter still stands therefore. Indeed it stands with increased emphasis, for without there having been biological evolution for the conditions of space, the observed situation would be implausibly fortuitous.

It is important to extend the brief discussion given by Arrhenius of the effects of ultraviolet light on bacteria and viruses. He was correct in his statement that the absence of oxygen would be helpful, especially with respect to the near-ultraviolet at wavelengths of about 3000 Å.[1] But the absence of oxygen is not sufficient to offset the damaging effect of ultraviolet at wavelengths of 2600 Å. Such radiation is injurious to genetic material, to DNA and RNA, changing the bonding of bases in such a way that they tend to stick together in pairs (especially when the base thymine is involved), instead of acting independently as in undamaged material.

If one were concerned only with the travel of bacteria within the solar system, and in gravitational orbits around the Sun, as we were in a recent book,[2] it would be possible to evade this ultraviolet difficulty in a simple way. If bacteria travelled in colonies, a quite thin surface for each colony would be sufficient to give protection to the bacteria inside, because ultraviolet light with a wavelength of 2600 Å penetrates only a small distance into organic material. But for the Arrhenius theory this argument cannot be used, since a colony of the required size, say 0.01 to 0.1 mm, would be too big for radiation pressure to have a significant effect on it.

There is a remarkable escape from this apparent dilemma, however, an escape unknown to Arrhenius. Bacteria possess enzymic repair mechanisms of remarkable subtlety against the damage done by ultraviolet light. Given ordinary light to work with, an enzyme is able to reverse the effects of the ultraviolet light with respect to the sticking of bases in pairs, returning them to independent operation, or in some cases snipping out a damaged structure and returning the bacterium to a viable condition.

The existence of this repair mechanism is another embarrassment to the Earth-bound theory of the origin of life. Bacteria on the Earth are not exposed to ultraviolet light with this wavelength of 2600 Å. The ultraviolet light, although present in

[1] 1 mm = 10 million Ångströms (Å).

[2] *Diseases from Space* (London: Dent, 1979).

the radiation from the Sun, is absorbed by a layer of ozone high in the Earth's atmosphere (30 km to 50 km altitude) and so does not reach ground level. Hence there can be no biological evolution to produce such a repair process, unless we recede more than 3.8 billion years into the remote past and suppose that the Earth did not then have an ozone layer in its atmosphere. Even so it would be remarkable for a complex mechanism to have survived over such a huge time-span without there being persistent evolutionary pressure to maintain it.

Nor is this the end of the problem for the Earth-bound theory. Phages are viruses that prey on bacteria. Phages are also damaged by ultraviolet light, and in much the same way as their hosts. But phages are too small in their genetic structures to be able to produce an enzymic repair mechanism of their own. Instead, they use the repair facilities of their bacterial hosts. This ability we must also suppose in the Earth-bound theory to have survived over an enormous time-span without there being evolutionary pressure to sustain it.

Sometimes a bacterial cell contains a number of viral particles of the same type. After damage by ultraviolet light it may happen that not one of the viral particles is able to restore itself to viability with the aid of the bacterial enzymes. What can happen now is that the several particles 'cannibalize'. Bits from the otherwise dead particles are fitted together to form one functioning composite particle.

All these properties are essential to survival, if bacteria and viruses are space-travellers, and especially can one understand the situation if highly resistant microorganisms studied in the laboratory have been space-travellers comparatively recently. We anticipate the discussion of a later chapter, by noting that viability in a virus is not a genetically unique property. There is not just one form of virus that will attack its host. Many variations will do so, as for instance there are many varieties of the common cold virus. After ultraviolet damage, followed by the cannibalizing process described above, it is very unlikely that the newly-assembled viable virus will be exactly like any one of its several progenitors. It will be a virus of the same broad type as before, but a different variety of that type.

We have just mentioned the many forms of the common cold virus. Many varieties occur also for the influenza virus. According to conventional medical ideas these new forms are supposed to arise spontaneously from the old as the viruses multiply within their hosts. Yet no such variations have been found when the viruses are caused to multiply under laboratory conditions. Indeed quite the reverse. Instead of developing new varieties with added virulence to their hosts, viruses passing through only a few generations in the laboratory become very mild, scarcely able to attack their hosts at all (see *Appendix* for a fuller discussion).

The variations of both the common cold and influenza viruses are indicative of repair after ultraviolet damage at 2600 Å. Since this kind of damage does not occur at ground-level, the implication is that these viruses are continuously irradiated either in space or high in the atmosphere above the ozone layer, more than 50 km above ground-level.

If bacteria and viruses were space-travellers they would need additionally to be resistant to X-rays, particularly to soft X-rays which stars like the Sun emit occasionally in bursts of great intensity, bursts which typically have less than an hour's duration. No such resistance would ever have been needed for life evolving at the surface of the Earth, because X-ray bursts from the Sun do not penetrate below an altitude of about 80 km. This must have been so for as long as the Earth has possessed an atmosphere, even in the earliest days of terrestrial life, and even if there was then no ozone layer. Hence, according to the Earth-bound theory of the origin of life, cells have never been routinely exposed to the intense bursts of soft X-rays which the Sun emits from time to time, and especially when sunspots are numerous.

Radioactive materials in the rocks and soil produce an effect on living cells much like that of X-rays, and it is true that life evolving at ground-level would always be exposed to this natural radioactivity. There is, however, an enormous difference of intensity between terrestrial radioactivity and X-ray bursts from the Sun. Damaging exposure to radiation, whether X-rays or radioactivity, is measured in terms of a unit known as the rem, standing for roentgen-equivalent-man. Terrestrial radioactivity causes an average exposure of about 0.1 rem *per year*. Radioactivity in the early history of the Earth must have produced a higher average exposure at ground-level, but not higher than about 1 rem per year, which is a minuscule rate compared to a big X-ray burst from the Sun, which would give an exposure in the vicinity of the Earth of about 1 rem *per second*. A single burst lasting for an hour would thus expose space-borne bacteria and viruses to a total radiation dose of about 3000 rem.

Here then we have a crucial difference. If life had originated on the Earth there would have been no sustained evolutionary pressure for cells to develop repair mechanisms against accumulated exposures of more than a few rem, and no sustained pressure at all for repair against sudden large doses. In this connection we note that some scientists have postulated very infrequent short episodes in the Earth's history when life at ground-level might have experienced high radiation doses, as for instance if a nearby star exploded with the production of a very large flux of cosmic-ray particles. But such incidents would be so brief, and so widely separated in time, that they could not maintain selection pressure. They could only influence biology in the negative sense of rendering a radiation-susceptible species extinct.

If, on the other hand, bacteria and viruses are space-travellers with the ability to reach a planet like the Earth moving around a star like the Sun then they must be capable of withstanding sudden radiation doses of many thousands of rem. The issue is clear-cut, and in essence it reveals the whole story. The facts are that lethal radiation doses for many bacteria exceed 100 000 rem, and in some cases exceed a million rem, while individual viruses can withstand doses of up to 10 million rem. These values relate to sudden exposures. We are not aware of data for slowly-accumulated doses, but tolerance to such doses is quite likely to be many orders of magnitude greater still. This data comes as near to supplying objective proof that bacteria and viruses are indeed space-travellers as one could hope to find.

The higher animals can also withstand far larger sudden radiation doses than evolutionary pressures at the Earth's surface could possibly have produced, demonstrating that the cells of higher animals have either been space-travellers themselves, or have been put together from components which have been space-travellers.

To destroy a culture of bacteria entirely it is necessary to kill every single bacterium, whereas to be lethal to a complex multi-celled animal only a moderate fraction of the constituent cells need to be destroyed. Hence it is natural to expect lethal radiation doses for complex animals to be significantly lower than for bacteria, without there being any particular difference in the resistance of the basic components. The facts accord with this expectation — 500 rem being a typical lethal dose for mammals, considerably less than for bacteria, but still many orders of magnitude higher than a completely Earth-bound evolution would have yielded. Man is resistant to radiation at a far higher level than the natural background at ground-level, which is why the dangers of civil nuclear-energy programs are minuscule. One would have to live for about 100 000 years in an all nuclear-energy economy to accumulate a man-made radiation dose of 500 rem, and even 500 rem received over very many years would do little harm. It is *sudden* doses of 500 rem that are lethal.

Returning now to Arrhenius, the properties of cells that we have just discussed can in our view be taken as proof that life, in its basic biochemistry, is of extraterrestrial origin. Certain of the details of the Arrhenius theory, but not its main principle, are in need of revision, however. Particularly, the last two points in the presentation of Chapter 2 must be changed. The changes turn out to add further strength to the theory, as we shall find in Chapters 5 and 6. We end this chapter by explaining why the two points in question need rediscussion.

A bacterium expelled from our solar system would be most unlikely to find another star system. Unless precisely directed at another star, radiation pressure from that star would not check the high speed of motion of 100 km s^{-1}, or perhaps more, of the bacterium. Radiation pressure could only check the component of the motion directed towards the star, and there would in general be an unchecked component at right angles to the line joining the star to the bacterium (radiation pressure cannot change the so-called angular momentum around the star). In other words, the bacterium would pass the other systems at high speed in a kind of sideways motion, except in the improbable case of a bullseye shot on a body in the other star system.

More particularly, however, such a bacterium would be most unlikely to reach any other planetary system. This is true even if we suppose most stars to have planets like our own solar system, because a small particle expelled by our system would be caught up by the interstellar gas before it found another system at all.

This second flaw in the argument of Arrhenius was not of his own making. Although the first evidence of a gas lying between the stars had been obtained by J. Hartmann in 1904, the concept of an interstellar gas had not yet become accepted by astronomers. Indeed just the reverse. It was strenuously maintained by most astronomers that there was no such all-pervasive medium. Even clear-cut examples to the contrary, like the well-known gaseous nebula in Orion shown in Fig. 4.1, were thought of as exceptional objects. It is estimated nowadays that within our galaxy there are some 4000 gas clouds with the mass and size of those in Fig. 4.1.

Figure 4.1
This nebula in the constellation of Orion contains sufficient material to form more than a hundred thousand new star systems. There are about 4000 such clouds of gas in our galaxy.

Figure 4.2 shows the great rift of the Milky Way in the constellation of Sagittarius. When *Worlds in the Making* was being written, this cleft was thought to be due to a real absence of stars. Today we know it to be caused by an absorbing cloud of material which lies between us and the stars beyond. The main cause of the absorption is not gas, however, but small particles known as 'grains', or sometimes as 'dust' — which are about the same general order of size as bacteria. While some bacteria are more or less spherical in shape — the bacterium causing pneumonia, for example — most are of rod-like forms, as are most of the grains responsible for the absorption seen in Fig. 4.1. Bacteria expelled by radiation pressure from the solar system would in all respects have properties exactly like these grains, not only in their physical size and shape but also in their optical properties. Whether or not the grains of Fig. 4.2 actually are bacteria is a question to which we shall return in Chapter 7.

The argument used by Arrhenius to explain the egress of cells from the Earth's atmosphere was highly ingenious, but we doubt that it would be effective. The problem is that, while convection currents do carry small particles high in the atmosphere, such rising currents do not go to the top as Arrhenius supposed. Convection currents normally stop at a height of about 15 km, with some 10% of the atmosphere still above. Volcanic eruptions throw small particles of dust to still greater heights from time to time, but even these do not reach the 'top'.

It will be recalled from Chapter 2 that Arrhenius was concerned only with overcoming gravity:

> 'We can, without difficulty, calculate the intensity of an electric field capable of urging the charged spore of diameter 0.00016 mm upward against the force of gravity ...'

The calculation and argument which followed presumed gravity to be the only restraining force, whereas for particles as small as 0.00016 mm the important force to be overcome is the viscous

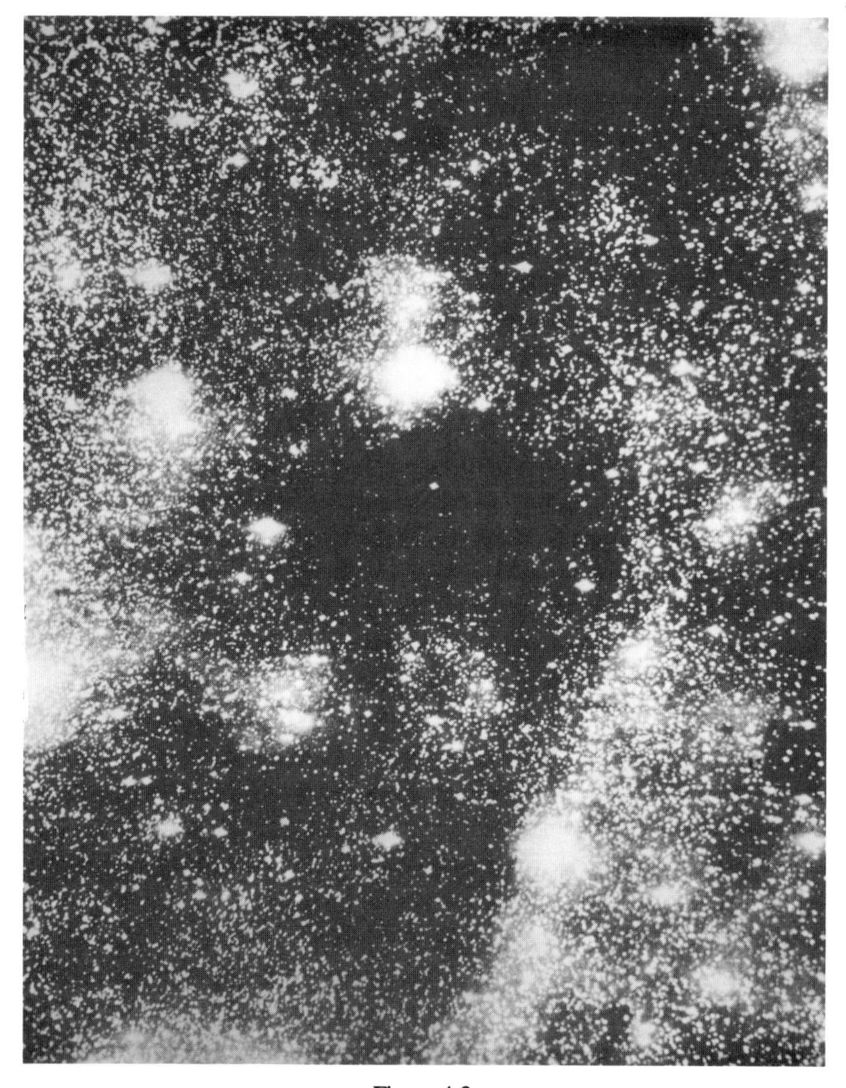

Figure 4.2
This dark cleft in the Milky Way is caused by a vast number of small particles with a similar size to bacteria. Most bacteria have elongated rod-like forms, as have these interstellar grains. (Courtesy of Hale Observatories.)

drag of the atmosphere on the particles. For particles the size of bacteria, the atmosphere, even at a height of 50 km, is sticky like treacle, and the particles do not gain egress through electric forces. For particles as small as viruses, electric forces are much more effective, except that the sense of such forces on very small particles is believed to be downward towards the Earth, the opposite from the sense required by Arrhenius.

Let us now summarize the position we have reached.

Bacteria and viruses are excellently equipped to make long journeys in space, but the beginnings and endings of their journeys cannot follow the details suggested by Arrhenius. Living cells are not transported directly from a planet in one star system to a planet in another star system. Beginnings and endings must be managed in a different way, which we shall consider in the next two chapters.

5
COMETS

BARNARD PICTURES OF HALLEY'S COMET

Taken at Yerkes Observatory May 4, They Tally with Observation from Times Tower May 5.

VIEWED BY MISS PROCTOR

Negatives Show the Tail Extending 20 Degrees, Equivalent to 24,000,000 Miles in Length.

IN COMET'S TAIL ON WEDNESDAY

European and American Astronomers Agree the Earth Will Not Suffer in the Passage.

TELL THE TIMES ABOUT IT

And of Proposed Observations—Yerkes Observatory to Use Balloons if the Weather's Cloudy.

TAIL 46,000,000 MILES LONG?

Scarfed in a Filmy Bit of It, We'll Whirl On In Our Dance Through Space, Unharmed, and, Most of Us, Unheeding.

SIX HOURS TO-NIGHT IN THE COMET'S TAIL

Few New Yorkers Likely to Know It by Ocular Demonstration, for It May Be Cloudy.

OUR MILLION-MILE JOURNEY

Takes Us Through 48 Trillion Cubic Miles of the Tail, Weighing All Told Half an Ounce!

BALLOON TRIP TO VIEW COMET.

Aeronaut Harmon Invites College Deans to Join Him in Ascension.

MAY SEE COMET TO-DAY.

Harvard Observers Think It May Be Visible in Afternoon.

MAY BE METEORIC SHOWERS.

Prof. Hall Doubts This, Though, but There's No Danger, Anyway.

YERKES OBSERVATORY READY.

Experts and a Battery of Cameras and Telescopes Already Prepared.

CHICAGO IS TERRIFIED.

Women Are Stopping Up Doors and Windows to Keep Out Cyanogen.

(Facsimile headlines from the **New York Times** coverage of Halley's comet on May 10, 16, and 18, 1910)

A typical comet has a mass of about 10 000 000 000 tonnes. From an everyday point of view this is a large chunk of material, and a comet hitting the Earth — as a bit of a comet hit Siberia in 1908 — can cause a great deal of local damage. But a typical comet is a tiny object compared to a planet like the Earth, which has a mass of about 6 000 000 000 000 000 000 000 (6×10^{21}) tonnes. There are very many comets, however. Some astronomers think there may be as many as one million million of them, totalling in mass more than the whole Earth. Collectively, all the comets of the solar system are equivalent in quantity of material to a substantial planet.

The material of which comets are made is lifelike in its basic composition. Table 5.1 shows a comparison between the elemental composition in life, comets and other cosmic systems recently published by Professor A.H. Delsemme. The close correspondence between the columns for comets and bacteria is quite remarkable. Thus if one were to take living material and to break it down into very simple molecules, the result would be very much

Table 5.1. Relative abundances of major elements
(courtesy A.H. Delsemme)

	Cosmic	Volatile fraction in comets	Bacteria	Mammals
Hydrogen	76.50	56	63.0	61.0
Oxygen	0.82	31	29.0	26.0
Carbon	0.34	10	6.4	10.5
Nitrogen	0.12	2.7	1.4	2.4
Sulfur	0.0015	0.3	0.06	0.13
Phosphorus	0.00002	—	0.12	0.13
Calcium	0.0002	—	—	0.23
H/O	14000	1.8	2.2	2.3
C/O	0.64	0.32	0.22	0.40
N/O	0.12	0.08	0.05	0.09

Figure 5.1
The comet Mřkos, 22 August 1957. (Courtesy of Hale Observatories.)

like the chemical composition of the gases which are exuded by comets.

Figure 5.1 shows the comet Mřkos. When a comet is photographed with very short exposure it appears as a star-like point of light. This is the region containing most of the comet's material, and from which gas and small particles are evaporated. As the exposure is progressively increased, the point of light grows into a structure known as the 'coma', from which a 'tail' is eventually found to emerge. The coma thus forms the head of the comet on its sunward side, with the tail streaming away from the Sun. The tail of Comet Mřkos can be seen to have two parts, a smooth shorter part on the right in Fig. 5.1 and longer wavy structures on the left. The latter are the gases with a chemical composition very like living material. The smooth part of the tail is made up of particles with the same general order of size as bacteria. These particles in Fig. 5.1 are being expelled from the solar system by solar radiation, the process discussed by Arrhenius. Just from the orientation of the dust tail in Fig. 5.1 one can deduce that the Sun is in a direction towards lower left.

The inorganic material at the Earth's surface from which life is supposed in the Earth-bound theory to have evolved, is not at all life-like in its chemical composition. Particularly, the ratios of the abundances of the important life-forming atoms — hydrogen, carbon, nitrogen, and oxygen — are quite different at the Earth's surface from their ratios in living material. In comets, however, the ratios are substantially the same. From a straightforward chemical point of view it is therefore preferable to seek the generation of life in the very large total quantity of cometary material than it is in the much smaller quantity of volatile materials here on the Earth.[1]

We have developed this point of view in some detail in two earlier books: *Lifecloud* (London: Dent, 1978); *Diseases from*

[1] For the crucial element carbon, the quantity in the comets is of the order of a million times greater than the quantity in carbonate rocks in the Earth's crust.

Space (London: Dent, 1979). We took the spontaneous generation of life to have occurred in comets, with living cells eventually becoming transported from the comets to the Earth. Here we are concerned with a different concept, however. The position we have just reached, in which comets play a more dominating role than planets, fits immediately into the picture of Arrhenius. We dispense with the weak part of the old picture, according to which life goes directly from a planet in one star system to a planet in another star system. Instead, we first have life going from comets in one star system out into the interstellar gas, particularly into clouds like the Orion nebula in Fig. 4.1, just as some of the particles in the dust tail of Fig. 5.1 are actually doing. When new star systems form in such clouds — as indeed new systems are forming all the time — the new systems receive the seeds of the life that are already present in the clouds. For those new systems that develop in the manner of our own solar system, with comets on the far outside where material remains cool, the seeds of life would not be destroyed — they would pass to the comets of the new system. Encountering material of just the right chemical composition, they would grow explosively in number. The comets of the new system would thus serve as *amplifiers* of the life from the old system.

Comets are both amplifiers and distributors. When, from time to time, a comet comes sufficiently close to its central star, evaporation of water and other volatile materials takes place, and with this evaporation the seeds of life — after being greatly amplified in number — come free into the tail of the comet (Fig.5.1). The seeds, repelled by radiation pressure, are then on their way outwards again, back to the interstellar clouds where they will join the seeds from many other star systems.

There is no requirement now for life to have emerged in a complete form in any one star system. Bits and pieces from many systems, different one to another in the physical details of their environments, contributed to an accumulation of biochemical components within the interstellar clouds. As new star systems condensed in the clouds, the various bits and pieces were tried out together in different ways in the different star systems, until at last a combination occurred that had the devastating amplification of numbers so characteristic of living cells — one cell makes two, two make four, and so on. Once such a self-amplifying biochemical system had been found it simply mopped up all the life-forming materials. For the comets of our own system there were probably several Earth masses of such materials, while for our galaxy as a whole there may well have been of the order of one million million Earth masses going into living forms, all with the same basic biochemistry.

The ability of bacteria and viruses to withstand large X-ray doses can now be assessed in its proper setting. Living cells travelling in space, even if not close to one particular star, are still exposed to X-rays and to cosmic rays (which act on them rather like radioactive particles). For living material, with its large chemical content of carbon, nitrogen, and oxygen, X-rays are considerably more damaging than cosmic rays. X-rays are contributed in part by a large number of ordinary stars like the Sun, in part by a much smaller number of specially intense X-ray stars, and in part by other galaxies, especially galaxies that are exceptionally strong X-ray sources, like the one shown in Fig. 5.2.

For an unshielded cell at a typical point of interstellar space the most deleterious effects come from soft X-rays. Calculation shows that a dose of a million rem due to soft X-rays would accumulate in about 10 000 years, which is about the time a cell moving at a speed of 100 km s^{-1} would need to cover the distance of three light years by which neighbouring stars are typically separated. For cells inside a cloud like the Orion nebula (Fig. 4.1) there would be shielding by the gas of the cloud against soft X-rays. Harder X-rays would still penetrate the cloud, however, but would need a million years to irradiate cells with a dose of a million rem. Thus inside the interstellar clouds, bacteria and viruses with the known ability to withstand radiation doses of the order of a million rem, would have

Figure 5.2
The galaxy M87 (NGC 4486) in the constellation of Virgo. This galaxy is one of the strongest known emitters of X-rays. (Courtesy Hale Observatories.)

quite long lifetimes. Indeed the lifetimes could well be considerably longer than these estimates might suggest, because X-ray doses accumulate only very slowly in space, away from the vicinities of stars, whereas lethal doses measured for viruses and bacteria in the laboratory are for sudden flashes of X-rays — and experience shows that sudden flashes at high intensity are far more damaging than slowly accumulated doses.

It is interesting to consider the conditions under which living cells could make the journey, not merely from star to star, but from one galaxy to another. The speeds with which particles of appropriate sizes in the tails of comets are expelled from the solar system are generally about 100 km s^{-1}, but higher speeds are certainly attainable. Such higher speeds arise for instance if particles ball up into a loose structure with an effective density perhaps of only one-tenth that of water, a situation which is actually believed to happen for some cometary particles. In sufficiently exceptional cases, speeds generated by radiation pressure of 1000 km s^{-1} might well arise, and a cell travelling at this speed could make the journey from our galaxy to the neighbouring large galaxy in the constellation of Andromeda (Fig. 5.3) in less than a billion years, in about one-fifth of the span of time for which life has existed here on the Earth.

The question of whether life could step from galaxy to galaxy is not so much a matter of time therefore as of whether cells could withstand exposure to X-rays for as long as a billion years. This turns again on the uncertain problem of the effects of low intensity irradiation experienced over long periods.

If indeed life can cross from one galaxy to another we could imagine a chain being set up, with life spreading galaxy by galaxy like a bucket handed along a chain. How far could such a step by step process extend? We know that exceptional violent explosions of entire galaxies as occurred in M82 could lead to the expulsion of vast quantities of grains (Fig. 5.4). If travel speeds of up to 10 000 km s^{-1} could occur in such outstanding situations, then over the age

Figure 5.3
A living cell expelled by our galaxy at a speed of 1000 km s^{-1} would reach the great galaxy in the constellation of Andromeda (shown here) in the span of time comparable to that which has elapsed since the late Precambrian, about 700 million years. (Courtesy Hale Observatories.)

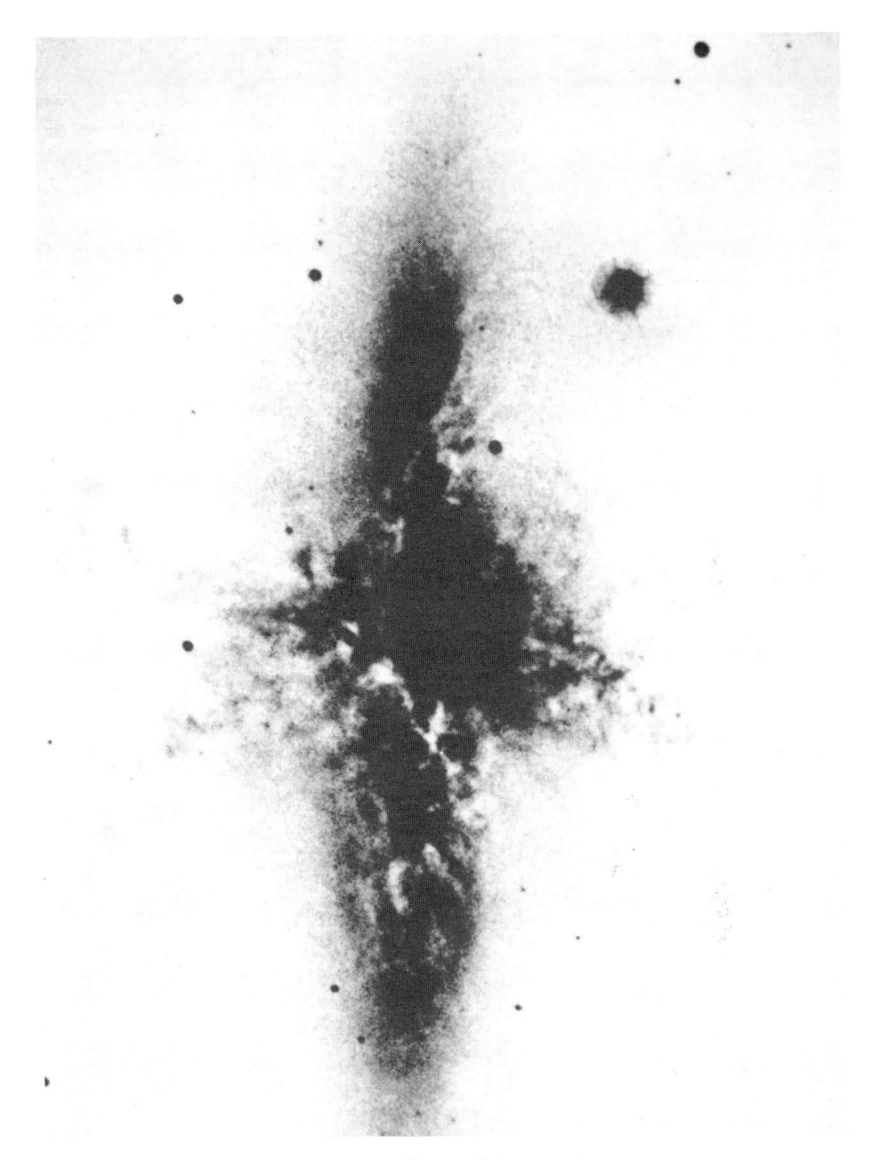

Figure 5.4
A violent explosion in the galaxy M82 has led to the expulsion of vast quantities of grains into intergalactic space.

of our own galaxy, 10 to 15 billion years, life could spread from its starting point to some 300 million light years. About a hundred thousand galaxies could be reached. Subject to the uncertain X-ray problem discussed above, we can therefore imagine biochemical unity becoming established through a fair-sized portion of the whole universe.

We have remarked on the complexity of the chemistry of life, which we now think unlikely to have arisen within the solar system, let alone on the Earth. Life may not have wholly arisen even in our galaxy. Possibly the resources of many galaxies were needed to build so intricate a biochemical system. Possibly we have relatives, not just in our own galaxy, but in quite a number of other galaxies among the nearer and brighter ones.[1]

[1] In a steady-state form of cosmology the spread of life could be much wider still.

6
COSMIC EVOLUTIONARY CYCLE

The concepts discussed in the previous chapter have a dramatic effect on ideas about the early days of the solar system. Instead of the need for the difficult, and necessarily slow, spontaneous generation of life there would have been explosive, almost instantaneous, amplification of life forms added to the outside of the solar system from the interstellar cloud in which our system condensed.

In the earliest days the primitive Sun was superluminous, so much so that the temperature at the Earth's distance from the Sun was about 1200 °C, like a blast furnace. At the distance of the outer planets, Uranus and Neptune, the temperatures measured on the absolute scale (Kelvin) was about five times lower, however, about 300 K (about 30 °C), ideal for life processes to operate. Given a vast mass of chemical foods, given just the right temperature, and given the initial presence of the seeds of life, the number of living cells must literally have increased to an astronomical level, perhaps to as many as 10^{40} such cells. The situation is reminiscent of the story which Ibn Khallikán relates in the Life of Abú Bakr al-Súli:

> 'I have met many people who believe al-Súli to have been the inventor of chess, but this is an erroneous opinion, that game having been imagined by Sissa the son of Dáhir, an Indian, for the amusement of King Shihrám.
>
> 'It is said that, when Sissa invented the game of chess and presented it to Shihrám, the King was struck with admiration and filled with joy; he had chess-boards placed in the temples, and he expressed his opinion that the game was the best thing man could study, inasmuch as it served as an introduction to the art of war, and that it was an honour to the Faith and the World, as well as the foundation of all justice.
>
> 'The King also manifested his gratitude and satisfaction for the favour which Heaven had granted him in shedding lustre on his reign by such an invention, and he said to Sissa, "Ask for whatever you desire".
>
> "I then demand," replied Sissa, "that a grain of wheat be placed in the first square of the chess-board, two in the second, and that the

numbers of grains be progressively doubled till the last square is reached: whatever this quantity of wheat may be, I ask that you bestow it on me." The King, who intended to make him a worthy present, exclaimed that such a recompense would be too little, and he reproached Sissa for asking so inadequate a reward. Sissa declared that he desired no other gift and, heedless of the King's remonstrances, he persisted in his request. The King at last consented, and gave orders that the required quantity of grain be given to him. When the chiefs of the royal house received their orders they calculated the amount, and answered that they did not possess near as much wheat as was required. When these words were reported to the King, he was unable to credit them and had the chiefs brought before him; when questioned on the subject they declared that all the wheat in the world would be insufficient to make up that quantity. They were asked to prove the truth of this contention, and by a series of multiplications and reckonings they demonstrated that such was indeed the case. The King then said to Sissa, "Your ingenuity in imagining such a request is even more admirable than your talent in inventing the game of chess".

'The way in which the doubling of the grains of wheat is to be done consists in the calculator placing one grain in the first square, two in the second, four in the third, eight in the fourth, and so on, until he comes to the last square, placing in each square double the number contained in the preceding one. I was doubtful of the contention that the final amount could be as great as was said, but when I met one of the accountants employed at Alexandria, I received from him a demonstration that convinced me that their statement was true: he placed before me a sheet of paper, on which he had calculated the amount up to the sixteenth square, obtaining the result 32,768. "Now," said he, "let us consider this number of grains to be the content of a pint measure, and this I know by experiment to be true" — these are the accountant's words, so let him bear the responsibility — "then let the pint be doubled in the seventeenth square, and so on progressively. In the twentieth square it will become a *wayba*, the *wayba* will then become an *ardeb*, and in the fortieth square we have 174,762 ardebs; let us consider this to be the contents of a corn-store, and no corn-store contains more than that; then, in the fiftieth square we shall have

> the contents of 1,024 stores; suppose these to be situated in one city — and no city can have more than that number of granaries or even so many — we shall then find that the number required for the sixty-fourth and last square corresponds to the contents of 16,384 cities; but you know that there is not in the whole world a greater number of cities than that. This demonstration is decisive and indubitable".'

This is just the way it is in biology, except that to yield 10^{40} grains of wheat, equalling the number of living cells that could have been produced in the early days of the solar system, the chess-board would need to have had 134 squares. Living cells multiply typically in generation times of two or three hours under terrestrial conditions. One hundred and thirty-four such generations would require a mere couple of weeks. Multiplication times at the outside of the solar system, inside comets or perhaps even in the gas phase, would be longer than this of course, but there is no reason why the main amplification in numbers should not have occurred in a few orbital periods around the Sun. A few centuries, or one or two thousands of years, seems the likely time-scale in which the biological explosion would take place.

The potentiality of living systems to increase enormously in their numbers is never given real scope to operate in orthodox biology. The potential appears in the theory of natural selection, but only in a minor degree. No creature is ever considered to multiply in number like the grains of wheat in the above story. In a personal notebook Darwin himself likened the potentiality for great expansion to the driving in of a wedge:

> 'On the average every species must have some number killed year with year by hawk, by cold, &c. — even one species of hawk decreasing in number must affect instantaneously all the rest. The final cause of all this wedging must be to sort out proper structure ... One may say there is a force like a hundred thousand wedges trying to force every kind of adapted structure into the gaps in the economy of nature, or rather forming gaps by thrusting out weaker ones.'

The concept is clearly of a jostling for advantage. Small gaps appear and are immediately filled by an adapted animal through its potentiality for rapid expansion. But these variations are only what mathematicians call small perturbations. Nowhere in orthodox biology is an enormous expansion contemplated, not into a mere 'gap' but into virgin territory, in the present case into the virgin territory provided by the formation of a new star system. The fact that orthodox biology provides no realization for the most crucial property of living cells shows in itself that something is seriously amiss with orthodox dogma.

Nuclear energy provides a quite close analogy to the biological situation. Nuclear energy can either be released explosively as in a military bomb or released in a controlled way as in a civil energy-producing reactor. Natural selection in biology, with all its ecological feedback mechanisms — as in the above quotation from Darwin — is much like the feedback systems used in the controlled release of nuclear energy. Any tendency of the energy-production to run away, or any tendency for an animal to increase too much in its number, is quickly prevented by a feedback mechanism. But there is nothing analogous in orthodox biology to the unrestrained release of nuclear energy occurring in a bomb. It is just this missing alternative that is provided here by the conditions in the early solar system. The biology of the outer regions of the solar system (with especial reference to the comets) provides the analogy to the bomb.

Astronomy draws on nuclear energy in both controlled and explosive ways, and it would be surprising if biology does not draw on the amplifying capacity of living cells both in controlled and explosive ways. Ordinary stars like the Sun generate nuclear energy in a stable process, whereas exceptional stars undergo explosions in conditions that are similar in many respects to a nuclear weapon. Figure 6.1 shows the debris from an explosion observed in 1054 AD. During the stable stages in the life of a massive star, the elements carbon, nitrogen, oxygen, sodium,

Figure 6.1
The Crab Nebula in the constellation of Taurus, debris from the supernova explosion observed in 1054 AD. (Courtesy of Hale Observatories.)

sulphur, chlorine, calcium, iron, and nickel[1] are produced from hydrogen and helium, whereas in explosions like that of Fig. 6.1 these higher elements are scattered into space, where they become available for incorporation into new stars and planets, and into life. Without the explosions of stars it is doubtful that life could exist.

Our view is that bacteria and viruses, and possibly more complex cells also, exploded in their numbers during the early history of the solar system, possibly within a few thousand years of the Sun being formed as a protostar. What would be the subsequent fate of the biological material so generated? Some would be swallowed into the planets Uranus and Neptune and be destroyed, but some would go into the great cloud of billions of comets which still surrounds the solar system. The hot primaeval Sun cooled off, and the resulting lowering of temperature to less than −100 °C for the material inside comets (on the outside of the solar system) caused living cells in the cometary material to become hard-frozen. Just as in the theory of Arrhenius, this deep-frozen life can persist essentially for ever, and it is still with us today.

After spending billions of years far out from the Sun, with its material hard-frozen in the great cold which exists beyond the planet Neptune, a comet may be deflected by the gravitational field of a passing star in such a way that it follows a new orbit that traverses the inner part of the solar system. For a particular comet this is an improbable happening, but there are so very many comets that it happens every year for two or three of them, as it has happened for Halley's comet, whose present orbit is shown in Fig. 6.2.

In a case such as Halley's comet, the frozen material tends to thaw at its surface during the passage through the inner regions closest to the Sun. Evaporation of gas and particles produces the coma and tail of the comet, as we saw in the previous chapter (Fig. 5.1). Radiation pressure now expels the particles at high speeds from the solar system, back to the clouds of gas (Fig. 4.1) that lie

[1] To name only the more abundant elements.

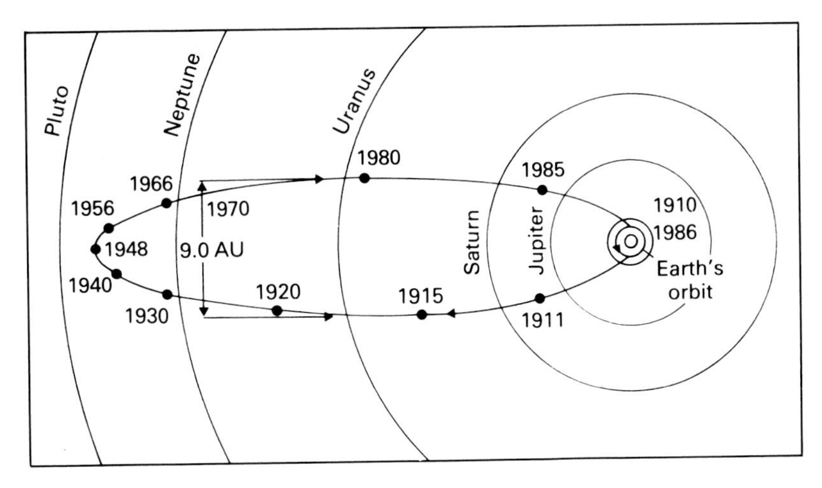

Figure 6.2
Present orbit of Halley's comet. Cometary objects with elongated orbits can bring material from the region of the outer planets into the region of the inner planets.

between the stars. The bacteria and viruses produced in the 'biological bomb' which occurred in the early days of the solar system are ejected out into space, thus completing the loop of Fig. 6.3.

Once a new star system condenses within an interstellar cloud, the right-hand side of Fig. 6.3 comes into operation very quickly. Suppose a mutation in a particular star system supplies an important biochemical improvement to some form of cell. The improvement is first stored away in a reservoir of billions of comets on the outside of the new system, and for the reasons discussed above it is then returned to the interstellar gas at a parsimonious rate of a few comets each year. The return may well start within a few thousands of years, which from an astronomical point of view is essentially instantaneous. The quickness of the first return permits the biological improvements to transfer almost immediately into further neighbouring star systems as they form, whereas the long drawn-out nature of the return gradually spreads the improvement

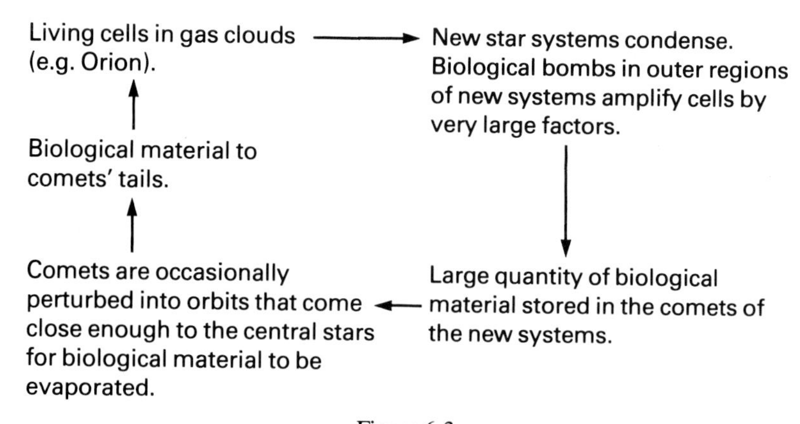

Figure 6.3
After being greatly amplified in quantity, life is returned to the interstellar clouds. Evolution in this cycle may have been the generator of life, starting from the simplest reproducible chemical systems.

over a considerable fraction of the galaxy. This is because stars move with respect to the interstellar clouds; and because of its motion the particular star system in which the biochemical improvement occurred would gradually seed many clouds quite distant from its own parent cloud.

The long-term storage inside comets prevents life in a galaxy from being wiped out by a sudden blast of X-rays, such as that which occurred in the centre of the galaxy M87 shown in Fig. 5.2. Likely enough, such a blast in our own galaxy would be lethal to cells within the interstellar clouds, but not to life hard-frozen within comets. So, as comets slowly returned their contents to the interstellar clouds, the loop of Fig. 6.3 would be re-established. Because there are very many star systems and because of the very long-term persistence of the left-hand arm of Fig. 6.3, it is essentially impossible, once life begins, ever to destroy it, except through a total burning out of all the stars in the galaxy.

Once life begins! This remains a crucial problem, but a problem that looks much more accessible here than it does in the very local

Earth-bound theory. We have the resources of our whole galaxy to draw on, with thousands of millions of star systems passing biochemical information between each other, very quickly on the scale of the distance between neighbouring stars, and on a longer time scale of billions of years over the whole of the galaxy.

For each star system, the loop of Fig. 6.3 contains on its right-hand side a biological explosion analogous to an enormous nuclear explosion. It is the nature of a biological explosion that not much in the way of constraints can be imposed on the non-lethal mutations which must arise inevitably during the explosion. The situation is just the opposite to that described in the earlier quotation from Darwin, in which a highly competitive 'wedging' is fiercely selective on mutations. The last stages of expansion, however, as the availability of suitable chemical materials decline, inevitably becomes competitive. The right-hand part of the loop of Fig. 6.3 therefore has both strong mutative and strong selective possibilities. The left-hand part of the loop is further selective of the properties which permit living cells to be space-travellers, properties like that of enzymic repair against radiation damage that we discussed in Chapter 4.

Without an enzymic repair process living cells are quickly destroyed by exposure to strong ultraviolet light at wavelengths around 2600 Å. This situation leads to a contradiction for the Earth-bound theory of the origin of life. Thus without ultraviolet light the enzymic property could never evolve, yet with solar ultraviolet light incident on a primitive chemical system life could not begin. Within the scheme of Fig. 6.3 this apparent paradox is resolved by life starting in star systems where there is no ultraviolet light to speak of, systems where the central star has a surface temperature no higher than 3000 K. Life broadcast into space from such low-temperature systems would then encounter ultraviolet light, but not at anything like the fierce intensity of ultraviolet from the Sun. There would, moreover, be opportunities if need be for the shielding of weakly-resistant life against even the comparatively feeble intensity of ultraviolet light in space, as for instance shielding inside gas clouds like the Orion nebula (Fig. 4.1). These are just the conditions needed for selection pressure to produce gradual evolution towards improved resistance to ultraviolet light (and also resistance towards X-rays). Our view is that the circulation of living cells around many millions of loops of Fig. 6.3 generated the persistent selection pressure which eventually produced the resistance to radiation damage that is now observed in actual cells.

With so powerful a system deploying the resources of the whole galaxy, one might be tempted to think that the problem of the spontaneous generation of life is close to solution. But issues both of detail and of principle remain. We defer this crucial conceptual problem preferring for the time being to be concerned with facts. We can ask, for example 'Is there observational evidence for the presence of biochemical materials in the interstellar clouds?' In the next chapter we show that there is indeed formidable evidence in support of this requirement of the above argument, and of the system of Fig. 6.3.

7
INTERSTELLAR BACTERIA

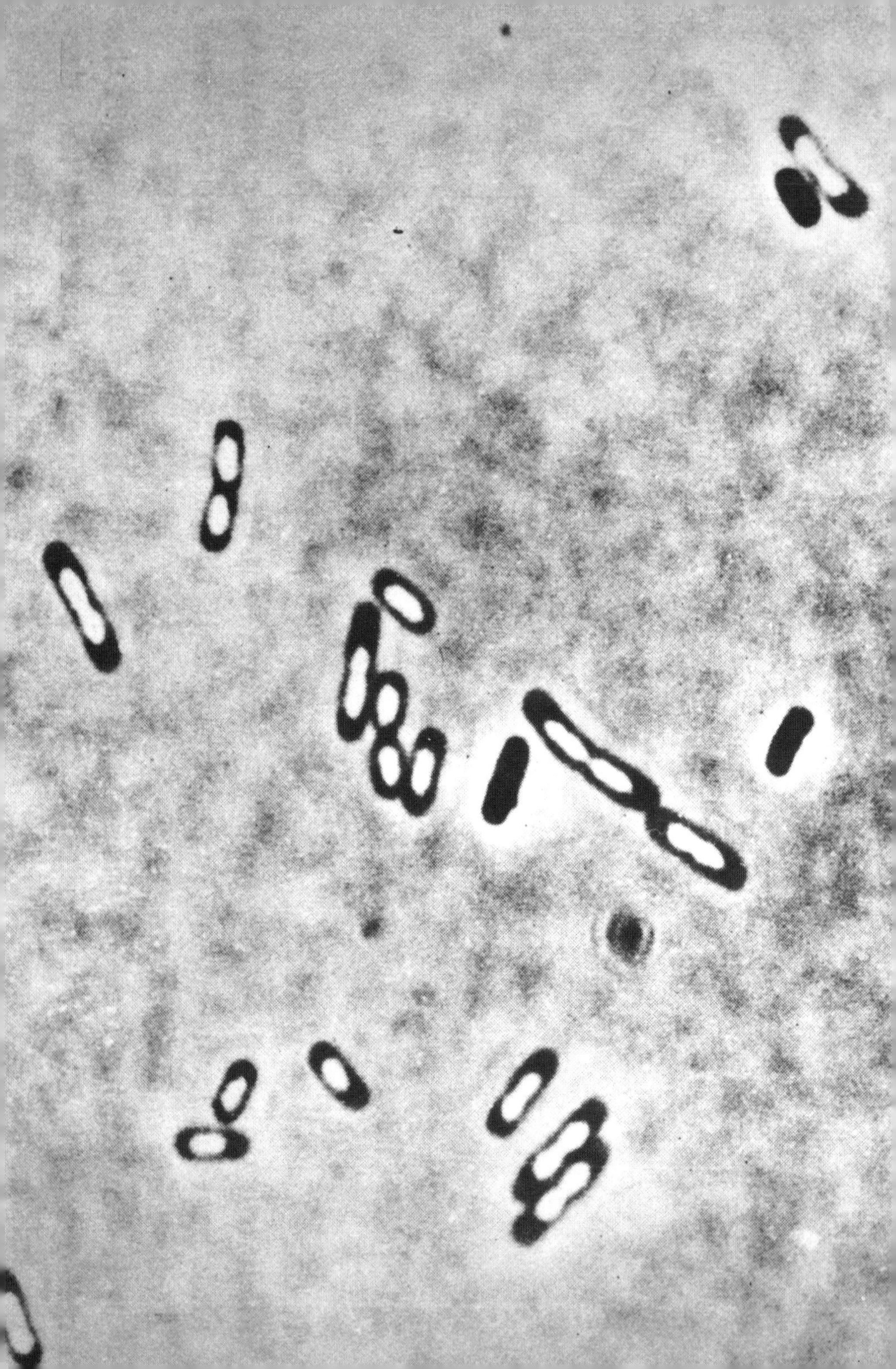

We have hinted in former chapters that bacteria might possibly be an important component of the interstellar dust, contributing to the blocking of starlight, which can be seen for example in Fig. 4.2. The amount of bacteria could be exceedingly large. With ten Earth masses from each of the estimated 200 000 million stars in our galaxy, the amount would be about 6 million times the mass of the Sun, which is quite comparable to the total quantity of all the interstellar grains in the galaxy.

Cellulose is an important component of the cell walls of plants and most animals, although bacteria have a substance glycon in place of cellulose. Glycon has heat emission and absorption properties very similar to cellulose, however, and these are uncannily like the observed properties of interstellar grains. The curve of Fig. 7.1 has been calculated for a heat source at a temperature of 157 °C, with the infrared radiation from the source passing through a quantity of perfectly dry cellulose particles.[1] The curve gives the expected heat energy transmitted through the cellulose at the various wavelengths specified on the horizontal scale. The points plotted in Fig. 7.1 are the observed heat intensities (given on the vertical scale) for the astronomical object with the catalogue designation OH26.5+0.6. The agreement between observation and calculation is good, and similar correspondences have been found for many other astronomical sources of infrared radiation.

There is additional strong confirmation from other astronomical observations that the grains are composed predominantly of carbonaceous materials. Thus if one were to exclude carbon, nitrogen, and oxygen from the grains there would be significantly too little solid material to explain the extent to which the grains are able to block starlight, as in Fig. 4.2.

To understand how this latter deduction is made, we must explain that astronomers have a quite accurate knowledge of the

[1] *Astrophys. Sp. Sc.*, 72 (1980), 247.

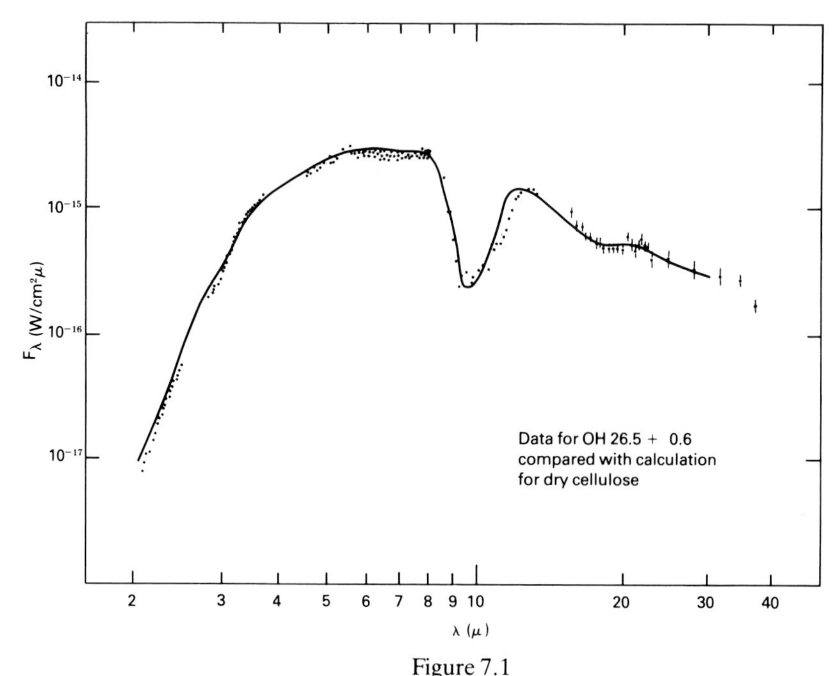

Figure 7.1
The ordinate is infrared flux (measured in W cm^{-2} per unit band width) and the abscissa is wavelength (in μm). The points are observations for the source OH 26.5 + 0.6, and the curve shows the effect of a cloud of anhydrous cellulose grains intervening between the observer and a more distant infrared source of temperature 430 K.

proportions of the various common elements as they exist on a cosmic scale, whether in the stars or in the interstellar gas and dust. Nowadays, the total quantity of interstellar material is also known, and hence the maximum amount of grains composed of magnesium silicates, of iron, and of other common grain-forming elements, but *excluding* carbon, nitrogen, and oxygen, can readily be calculated. It is this amount that is substantially too small to explain the observed blocking of starlight. But when carbon, nitrogen, and oxygen are taken to be formed into grains, as for instance into grains of cellulose or into bacteria, the amount of grains agrees with the amount needed to produce the observed blocking of starlight.

If most of the interstellar carbon and nitrogen is formed into grains then of necessity little of these elements can be left over in the interstellar gas, an expectation that has been confirmed by observation.

Already in Chapter 4 we noted evidence (from the so-called polarization properties of the interstellar dust) that the grains tend to be rod-like in their structures, as also are many types of bacteria. Furthermore, there is an observed difference between the blocking effect of grains on red light and on blue light which is just what would be expected if the grains were predominantly bacteria. This matter is so important that we will consider it in a little detail. The first overwhelming point concerns the sizes of interstellar grains. It has been known for twenty years or more that the particle diameters which contribute to the bulk of the extinction of visual starlight must be quite sharply concentrated at about 0.7 μm,[1] a requirement that defies explanation for grains of inorganic composition such as water-ice. Thus the sequence of evaporation and condensation phases to which water-ice grains would inevitably be exposed (from considerable variations in their temperature) must in general lead to a broad distribution of sizes,[2] not to a distribution concentrated near 0.7 μm.

Figure 7.2 gives the size distribution of a particular class of bacteria, those which form spores. Data of similar quality for other classes of bacteria do not seem to exist, but we understand from our enquiries that the situation is not expected to be much different from Fig. 7.2. The concentration at about 0.7 μm shown by the peak of Fig. 7.2 already gives strong support to the identification of the interstellar grains with bacteria.

[1] This takes account of a recent verification of an older assumption, that the grains scatter visual light more strongly than they absorb it. This property implies that the grains consist mainly of electrically insulating material. It is a property that rules out solid particles of carbon as the main constituent, although particles of solid carbon are almost surely present as a secondary constituent.

[2] By 'size' we mean diameter for spherical grains, and rod diameter for rod-shaped grains.

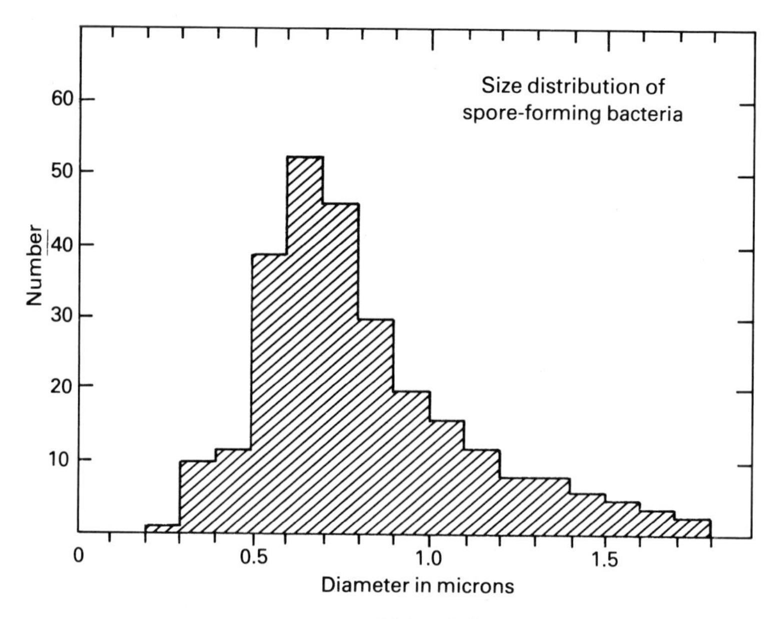

Figure 7.2
The size distribution of species of spore-forming bacteria.

Under terrestrial conditions about 70% of the interior volume of a bacterium is liquid water.[1] Since water passes comparatively freely through the cell membrane and out through the wall, it is experimentally possible to remove water progressively. The cell wall, being exceedingly tough, remains fixed and a cavity of increasing size then develops within the bacterium, as is illustrated in Fig. 7.3. It is important to decide how large a cavity develops under interstellar conditions, since the answer to this question affects the light scattering properties of the particle.

Carbon and oxygen atoms are about equally abundant in biological material. Oxygen atoms are nearly twice as abundant as carbon atoms in cosmic material, however. Since we know that most of the interstellar carbon is tied up in grains, it follows from these two facts, and provided the grains are bacteria, that one

[1] This is the fraction of water measured for the bacterium *E. coli*.

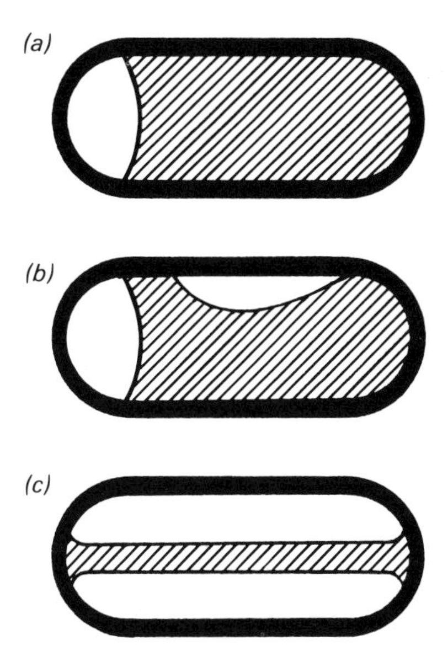

Schematic representation of three different degrees of plasmolysis in a rod-shaped bacterium: *(a)* slight; *(b)* extensive; *(c)* severe.

Figure 7.3
Three different stages in the removal of water from the interior of a rod-shaped bacterium.

oxygen atom is left over for every oxygen atom present in the biological material. This remaining oxygen atom is likely to be combined partly with metal atoms and partly with hydrogen as water, which at the generally low temperature of the interstellar clouds is likely to be condensed as ice within the bacteria. It is then an easy calculation to show that about 30% of the volume of an interstellar bacterium would be biological, about 5% would be ice, with the remaining 65% space. For such a situation one can demonstrate that a cloud bacterium would behave with respect to the scattering of light like a uniform particle of unusually low 'refractive index'. For water alone the refractive index is about 1.33. For biological material alone it is about 1.50. The effect of the 65% space within a bacterium would be to lower its effective

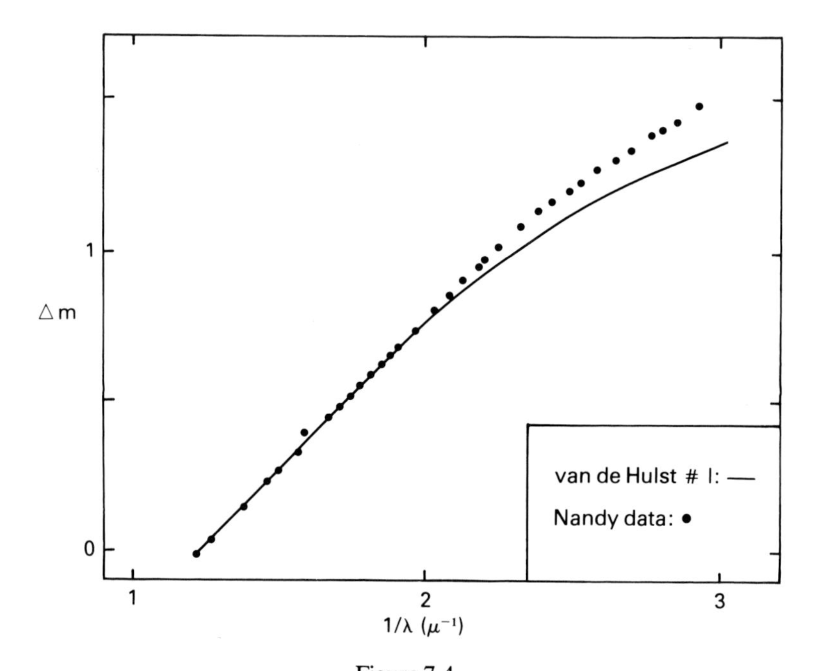

Figure 7.4
This is the best fit that can be achieved between the data points and grains with the refractive index of water. The left-hand scale is a measure of the scattering efficiency of the grains (in magnitudes) and the bottom scale is the reciprocal of the wavelength of the light (wavelength in μm).

refractive index to about 1.16. The crucial effect of this lowering of refractive index is shown from the comparison of Figs. 7.4. and 7.5, a comparison that is in itself essentially decisive. The very small discrepancies near $\lambda^{-1} = 2.3\ \mu m^{-1}$ which can be seen in Fig. 7.5 are due to the neglect in the calculations of a small amount of absorption of blue light produced typically by biological pigments.

We turn now to astronomical data at shorter wavelengths than those shown in Figs. 7.4 and 7.5 (wavelengths less than 0.3 μm). There is now the crucial difference that much of the blocking effect of the grains comes from true absorption, particularly over the

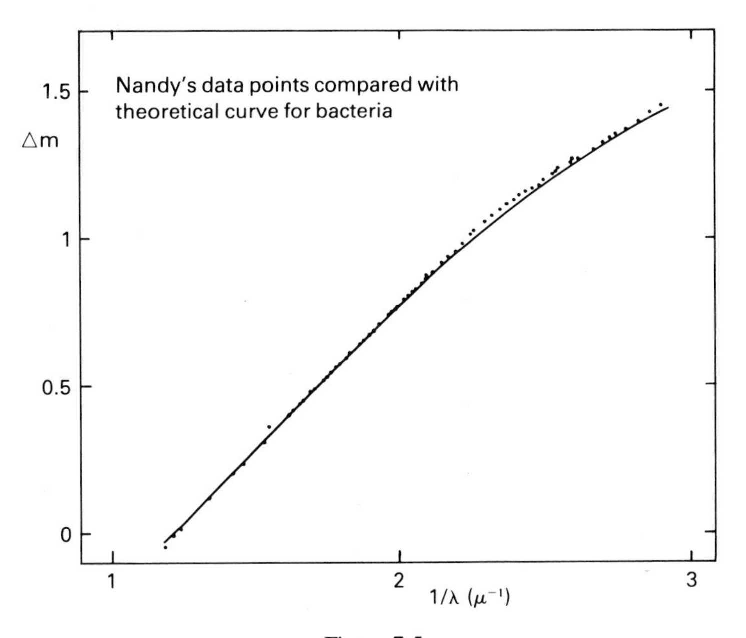

Figure 7.5
This is the fit that can be achieved when the refractive index is lowered due to the grains being hollow, as bacteria would be. The scales are as Fig. 7.4.

wavelength range from 0.3–0.18 μm, not simply from a scattering of light on its journey from a star to the Earth. The degradation of biological material into a carbonaceous soot causes strong absorption at 0.22 μm, precisely where the interstellar grains have their absorption maximum.

The observational data from several sources are shown by the points in Fig. 7.6, in which the left-hand scale is again a logarithmic measure of the fogging of starlight by the grains, and the lower scale is again the reciprocal of the wavelength expressed in micrometers (for a wavelength of 0.22 μm, the reciprocal is 4.55, corresponding to the peak of the figure). The solid curve in Fig. 7.6. shows the calculated behaviour of a mixture of particles comprised of rod-shaped bacteria under conditions we have just

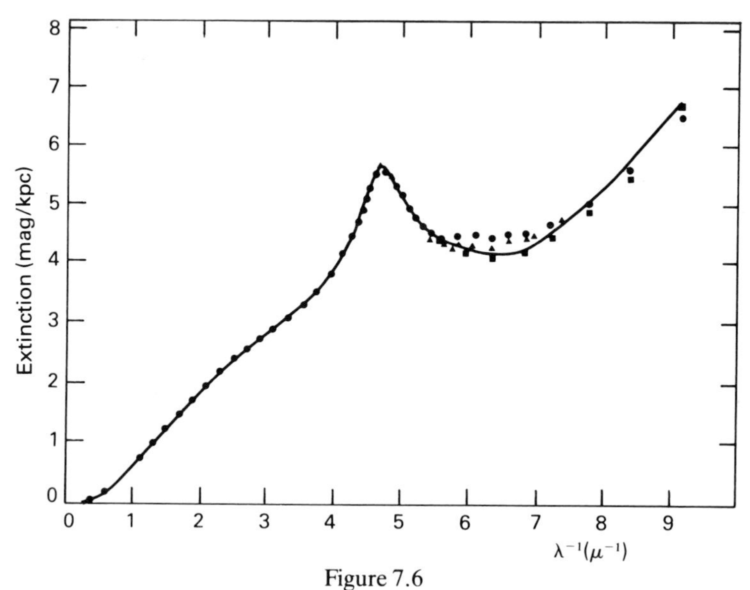

Figure 7.6
This is the best fit that can be achieved between the data points, and a mixture of bacilli (65%), mycoplasmas (25%) and graphite spheres (10%). The scales are in the same units as Fig. 7.4

described making up 65% of the mass, dielectric particles of organic composition of radii 0.04 μm making up 25% of the mass and graphite spheres of radii 0.02 μm making up 10% of the mass of dust.[1] The two latter components as we shall see next are also conected with the behaviour of cosmic microbiology.

A fraction of interstellar bacteria must be exposed to various forms of radiation that cause them to degenerate. Under harsh conditions some bacteria take refuge in the production of small, wall-less forms resembling the so-called mycoplasma cells, with sizes about one-fifth to one-tenth of normal bacteria. Instead of scattering light strongly in the wavelength region of blue visible light, such smaller cells would scatter most strongly for light with wavelengths around 0.1 μm. This effect can be seen in the

[1] Details of this calculation and comparison are published by us in *Evolution from Space* (London: Dent, 1981).

astronomical data, and in our model calculation it is represented by the properties of small dielectric spheres of radii 0.04 μm. The main effect of these wall-less cells, making up about 25% of the mass, is one of scattering not absorption, and this too is confirmed by observation at the shortest wavelengths to the right of the peak of Fig. 7.6. In the less protected regions of interstellar space a slow but inevitable degradation of these mycoplasma cells occurs under conditions where oxygen is essentially absent. The sequence of chemical changes that results will be similar to the processes by which coal and graphite form from terrestrial biological deposits. We would in this way obtain a population of spherical graphitic particles as required for the agreement shown in Fig. 7.6.

It is remarkable that all the diverse properties of the interstellar grains become immediately explicable in terms of bacteria. Unlike previous unsuccessful theories, no *ad hoc* mixture of several widely different kinds of particle is required. The different components discussed above all follow from biology itself.

A further confirmation, if need be, of the bacterial hypothesis comes from the fact that the observed scattering and absorbing properties of the interstellar grains have remarkable uniformity over the whole of our galaxy. Indeed this uniformity has recently been found to extend to the Magellanic Clouds, which are systems outside our galaxy. This implies strongly that grains have everywhere the same chemical composition and the same closely determined size distribution, an impossible condition to meet, we think, for particles of a non-biogenic character. It would surely be hard to conceive of astronomical proofs any stronger than these are found to be.

The facts of astronomy point strongly to interstellar space being chock-a-block with biological material. The evidence for this statement has not been invented to fit the biological theory of earlier chapters, because the evidence was discovered before its biological interpretation became apparent — much of it long before it could be understood.

8
THE PROOF OF A THEORY

DIALOGO
di
GALILEO GALILEI LINCEO
AL SER.mo FERD. II. GRAN. DUCA DI
TOSCANA

Scientists are uncomfortable with theories that stand in isolation, without connections to other aspects of science. One might wonder if this prejudice does not impede progress by turning attention always away from the unusual. To some extent this might be true. Nevertheless, experience shows that correct ideas normally do have many connections to other parts of science. Although to begin with new ideas may appear separated from all else, unsuspected relationships soon emerge. A correct idea is like a runner plant, sending out its roots in all directions, so that before long it becomes integrated with more familiar parts of science.

Experience tends to show that ideas which continue to stand in isolation are usually wrong. This circumstance makes it possible for one to go through life automatically disbelieving everything that is radically new, and indeed the instant rejection of every new idea is often thought meritorious and is said to be an expression of a 'healthy scepticism'. The saving grace is that the really good new things mostly assert themselves quite quickly in a form which even the healthy sceptic cannot deny. On the whole the scientific world makes few mistakes, but when it does so the mistakes can be very big indeed. Yet even an enormous communal error has little in the way of unpleasant consequences for the healthy sceptic, since by the circumstances of the case almost everybody else has made the same mistake, which can then be represented to the public, not as a gross error, but as an example of the profundity of nature.

In an essay published in 1971[1] Gunther S. Stent gives an interesting discussion of one of modern biology's big mistakes.

> 'The fantastically rapid progress of molecular genetics from about 1965 eventually obliged merely middle-aged participants in its early development to look back on their early work from a depth of historical perspective that for scientific specialities flowering in earlier times came only after all the witnesses of the first blossoming were long dead. It was as if the late-eighteenth-century colleagues

[1] *Paradoxes of Progress* (San Francisco: W.H. Freeman, 1978), Chapter 5.

of Joseph Priestley and Antoine Lavoisier had still been active in chemical research and teaching in the 1930s, after atomic structure and the nature of the chemical bond had been revealed. This somewhat depressing personal vantage provided a singular opportunity to assay the evolution of a scientific field. In reflecting on the history of molecular genetics from my own experience I found that two of its most famous incidents — Oswald Avery's identification of DNA as the active principle in bacterial transformation and hence as genetic material, and Watson and Crick's discovery of the DNA double helix — illuminate two general problems of cultural history. The case of Avery throws light on the question of whether it is meaningful or merely tautological to say that a discovery is 'ahead of its time', or premature ...

'My *prima facie* reason for saying Avery's discovery was premature is that it was not appreciated in its day. By lack of appreciation I do not mean that Avery's discovery went unnoticed, or even that it was not considered important. What I do mean is that geneticists did not seem able to do much with it or to build on it. That is, in its day Avery's discovery had virtually no effect on the general discourse of genetics.

'This statement can be readily supported by an examination of the scientific literature. For example, a convincing demonstration of the lack of appreciation of Avery's discovery is provided by the 1950 Golden Jubilee of Genetics Symposium, *Genetics in the 20th Century*. In the proceedings of that Symposium some of the most eminent geneticists published essays that surveyed the progress of the first 50 years of genetics and assessed its status at that time. Only one of the 26 essayists saw fit to make more than a passing reference to Avery's discovery, then six years old ... The then leading philosopher of the gene, H.J. Muller, contributed an essay on the nature of the gene that mentions neither Avery nor DNA.'

Now Dr. Stent, in the somewhat more technical discussion quoted below, goes on to describe the big mistake:

'Why could Avery's discovery not be connected with canonical knowledge? Ever since DNA had been discovered in the cell nucleus by Miescher in 1869 it had been suspected of exerting some function in hereditary processes. This suspicion became stronger in the 1920s, when it was found that DNA is a major component of the chromosomes. The then current view of the molecular nature of DNA, however, made it well-nigh inconceivable that DNA could be the carrier of hereditary information. First, until well into the 1930s DNA was generally thought to be merely a tetranucleotide composed of one unit each of adenylic, guanylic, thymidylic, and cytidylic acids. Second, even when it was first realised by the early 1940s that the molecular weight of DNA is actually much higher than the tetranucleotide hypothesis required, it was still widely believed that the tetranucleotide was the basic repeating unit of the large DNA polymer in which the four units mentioned recur in regular sequence. DNA was therefore viewed as a uniform macromolecule that, like other monotonous polymers such as starch or cellulose, is always the same, no matter what its biological source. The ubiquitous presence of DNA in the chromosomes was therefore generally explained in purely physiological or structural terms. It was usually to the chromosomal protein that the informational role of the genes had been assigned, since the great differences in the specificity of structure that exist between various proteins in the same organism, or between similar proteins in different organisms, had been appreciated since the beginning of the century. The conceptual difficulty of assigning the genetic role to DNA had not escaped Avery. In the conclusion of his paper he stated: "If the results of the present study of the transforming principle are confirmed, then nucleic acids must be regarded as possessing biological specificity the chemical basis of which is as yet undetermined."

'By 1950, however, the tetranucleotide hypothesis had been overthrown, thanks largely to the work of Erwin Chargaff. He showed that, contrary to the demands of that hypothesis, the four nucleotides are not necessarily present in DNA in equal proportions. He found, furthermore, that the exact nucleotide composition of DNA differs according to its biological source, suggesting that DNA might not be a monotonous polymer after all. And so when two years later, in 1952, Alfred Hershey and Martha Chase of the Carnegie Institution's laboratory in Cold Spring Harbor, N.Y., showed that on infection of the host bacterium by a bacterial virus at least 80% of the viral DNA enters the cell and at least 80% of the viral protein remains outside, it was possible to connect their conclusion that DNA is the genetic material with canonical knowledge. Avery's "as yet undetermined chemical basis of the biological specificity of nucleic acids" could now be seen as the precise sequence of the four nucleotides along the polynucleotide chain. The general impact of the Hershey–Chase experiment was immediate and dramatic. DNA was suddenly in and protein was out, as far as thinking about the nature of the gene was concerned. Within a few months there arose the first speculations about the genetic code, and Watson and Crick were inspired to set out to discover the structure of DNA.'

This discussion makes clear the trouble with the Arrhenius theory. It was disconnected from essentially every other aspect of science. Even the ability of seeds to withstand extreme cold (− 252°), which might have provided a link with experiment in the laboratory, was dismissed by Arrhenius as an inevitable consequence of the slowing down of chemical reactions at very low temperatures — although there is really much more to this property of living cells, as we pointed out in Chapter 4. Arrhenius himself remarks in the last two paragraphs of his book:

'All these conclusions are in beautiful harmony with the general properties of life on our earth. It cannot be denied that this interpretation ... is distinguished by perfect consistency, which is the most important criterion of the probability of a cosmological theory.

'There is little probability, though, of our ever being able to demonstrate the correctness of this view by an examination of seeds falling down upon the earth. For the number of germs which reach us from other worlds will be extremely limited — not more, perhaps, than a few within a year all over the earth's surface; and those, moreover, will presumably strongly resemble the single-celled spores with which the winds play in our atmosphere. It would be difficult, if not impossible, to prove the celestial origin of any such germs if they should be found ...'

Theories that are inherently not subject to test are quickly ignored, because science is really a study of the universe as we observe it to be, not as we think it to be. It is not surprising therefore that the Arrhenius theory has had little influence on astronomy and biology. From time to time it is mentioned in passing, usually without appreciation of its intellectual qualities or of the care that Arrhenius had taken over its details.

If Arrhenius had seized on the exposure of bacteria to ultraviolet light as a central problem, and had caused the astonishing enzymic repair mechanism (discussed in Chapter 4) to be discovered in the laboratory, it is likely that his theory would have become a matter of orthodox belief. As it was, the ideas of A.I. Oparin and J.B.S. Haldane passed into orthodox belief, for the reason that they led to laboratory experiments which apparently confirmed the correctness of the starting point of the Oparin–Haldane theory.

When high-grade energy, electrical or ultraviolet light, is injected into a mixture of simple chemicals, the molecules will be partially dissociated into their constituent atoms. Provided there is plenty of hydrogen, carbon, nitrogen, and oxygen in the mixture to start with, the resulting atoms of these elements will in some degree associate themselves into organic molecules as they recombine, although most of the atoms will go back into the simple inorganic molecules from which they started, and the concentrations of the organic molecules will decrease sharply with increasing complexity. Nothing approaching the complexity of cellulose, chlorophyll, or of a long chain protein will be obtained, although amino acids and simple sugars can be expected in some degree. If experiments are ingeniously arranged to drain off the amino acids and sugars as they are produced, in such a way as to protect them from being destroyed in their turn by the high-grade energy, the system can be made into an on-going producer of simple organic materials, which can then be said to form an 'organic soup'.

To the modern chemist or physicist, used to carrying through 'network' calculations with the aid of fast computers, it would be surprising if the experiments had revealed anything different. But when in the 1950s such experiments were actually made, notably by H.C. Urey and S.L. Miller, they caused something of a sensation and were claimed as a demonstration of the correctness of the Oparin–Haldane theory. Obviously they did no such thing. For life to have originated by spontaneous generation on the Earth, it was still necessary to show:

(a) How the simple sugars and amino acids of the dilute organic soup could become organized into vastly more complex biomolecules of highly specific kinds;
(b) Why the sugars and amino acids were not quickly oxidized;
(c) If there were no free oxygen, and consequently no ozone layer in the atmosphere, why the sugars and amino acids were not destroyed by ultraviolet light.

These further requirements of the terrestrial theory are overwhelmingly the major part of its problems.

There is still a more important reason, however, why the experiments are no proof of a terrestrial origin of life, and this is because the basic physics on which they rely must necessarily have universal applicability. The production of the simple organic building blocks of life — amino acids, nucleic bases, and sugars — using precisely the same method, is vastly more important on a universal than on a terrestrial scale. In our galaxy, a large fraction of the whole of the ultraviolet light of some 10^{11} stars in available, not just the ultraviolet light intercepted by the small area of the Earth from the Sun alone. If one considers the astronomical experiments to be relevant to the Oparin-Haldane theory, then they are of the order of 10^{20} times more relevant as the starting point of the scheme of Fig. 6.3.

Although every argument of chemistry, biology, and astronomy favours the scheme of Fig. 6.3, the theory would still be in psychological trouble if it remained within the straight jacket of the final paragraph of *Worlds in the Making* (quoted on p.90). But it is indicative of the correctness of the theory that this negative conclusion of Arrhenius falls by the wayside. The number of germs arriving at the Earth from outside is not 'extremely limited'. It is estimated that upwards of 1000 tonnes of cometary debris, the kind of debris we see in the dust tails of comets (Fig. 5.1), enter the Earth's atmosphere each year. If an appreciable fraction of cometary material consists of bacteria, this is sufficient to give an incidence each year of 10^{21} bacteria, an enormous number, while the number of viral particles could be much larger still. So far from it being 'difficult, if not impossible, to prove the celestial origin of ... germs', one can rather say that it would be impossible to avoid proving their existence once we look at the evidence. This we shall do in the next three chapters.

In Chapter 9 we examine the technical problem of the entry of bacteria into the terrestrial atmosphere. The speeds of entry are high. Even for clumps of bacteria that are too large to have been much accelerated by radiation pressure, the speeds range from about 10 km s^{-1} on the low side to about 70 km s^{-1} on the high side, and one can wonder if the sudden heating generated on entry into the atmosphere at such speeds might not have a lethal effect on living cells. We shall consider this problem on the basis that loose clumps of bacteria would immediately be separated by the sudden pressure of the atmospheric gas, and that the heating problem is therefore one for individual bacteria to cope with.

9

ENTRY OF BACTERIA INTO THE EARTH'S ATMOSPHERE

Because individual bacteria and viruses are affected deleteriously by ultraviolet light and X-rays from the Sun, we consider here the more favourable cases of bacteria that travel either in protective clumps or those which have only recently quitted their parent comets. In either case, radiation pressure will not be a serious factor in changing the motions, which will be those of particles moving in highly eccentric orbits around the Sun, as for instance the orbit shown in Fig. 6.2.

This restriction does not imply any denial of the radiation pressure argument of Arrhenius. Bacteria as they come free by evaporation from cometary material must present a wide range of cases, depending on the degree to which they are clumped, on the precise position at which they are released in their orbits, and on their own individual sizes. Here we are choosing from this wide range the narrower range that is best suited to survive entry into the terrestrial atmosphere.[1]

We consider any loose clumping of bacteria to be immediately destroyed on entry into the atmosphere, so that from there on each individual bacterium has to fend for itself. Just as with the return of astronauts, so it is best for bacteria to enter the atmosphere at glancing angles. For a spherical particle with the density of water, and for the most favourable angle of entry, there is a relation between the speed of entry, V say, the diameter d of the particle, and the temperature T to which the particle is heated as it is slowed down by the atmospheric gases.

A typical choice for V, representing the bulk of cometary particles in eccentric orbits around the Sun, is about 40 km s^{-1}. For spherical particles with this speed of entry into the atmosphere, the

[1] These are rather obviously not the cases where the bacteria are expelled by radiation pressure from the vicinity of the Earth, or those cases where radiation pressure generates higher speeds that would be lethal on entry into the terrestrial atmosphere, but which are not lethal for passage through the much more diffuse interstellar gas.

relation between d and T in degrees Kelvin is[1]

$$d = 1.52 \times 10^{-15} T^4 \text{ cm}$$

We have now to decide on a maximum temperature up to which living cells could survive for a time of some tens of seconds. Laboratory sterilization procedures under dry conditions involve heating to about 425 K for an hour or more. For a sudden flash heating of only ten seconds or so, a survival limit of $T = 500$ K would seem reasonable. Putting $T = 500$ K in the above equation gives $d = 0.95$ μm. For the entry speed $V = 40$ km s^{-1}, this is the largest diameter of a surviving spherical bacterium. In the case of a rod-shaped bacterium, the maximum rod diameter is less than this by about two thirds, i.e. $d = 0.6$ μm or thereabouts, but with the length of the rod being free.[2] A rod-shaped bacterium is not limited in its length by entry requirements into the atmosphere.

In Table 9.1 we give the dimensions of some well-known bacteria.

Table 9.1. The shapes and sizes of some bacteria

1	Micrococci		
	Staphylococci Streptococci Pneumococci	Each cell is approximately a sphere with a diameter of 1 μm.	
2	Baccili (rods)		
	Mycobacterium tuberculosis	0.3 μm diameter,	3 μm length
	Escherichia coli	0.5 μm diameter,	3–5 μm length
	Clostridium tetani	0.5 μm diameter,	5 μm length
	Pasteurella pestis	0.7 μm diameter,	1.5 μm length
	Brucella abortus	0.4 μm diameter,	0.6 μm length
3	Vibrios (wavy shape)		
	Vibrio cholerae	0.5 μm diameter,	2–3 μm length

[1] For details of the calculation leading to this equation see Appendix 2 of *Diseases from Space* (London: Dent, 1979).
[2] *Ibid.*

Less typical cases can occur when the plane of the Earth's orbit around the Sun and that of the comet happen to coincide. In the further special situation in which the closest distance to which cometary particles approach the Sun happens to be just equal to the radius of the Earth's orbit, the encounter speed between a cometary particle and the Earth is then either about 10 km s^{-1} or about 70 km s^{-1}, according to whether the cometary particle goes around the Sun in the same sense as the Earth or in the opposite sense. The speed of 10 km s^{-1} is the least choice for V, and 70 km s^{-1} is the largest choice when radiation pressure is considered unimportant.

A cometary particle accelerates as it approaches the Earth, due to terrestrial gravity, and it might be thought that an allowance should be made for this additional effect. But if we are seeking the smallest possible value of V, in order to estimate the largest living cell which could survive passage from a comet to the Earth, we must take account of the possibility that particles can be captured in more than one entry into the atmosphere. The first encounter has the effect of making the particle a satellite of the Earth, with an orbit that is then degraded gradually in further passages through the atmosphere. In such cases, $V = 10$ km s^{-1} is a correct estimate, leading for a spherical particle to the following relation between the diameter d and the temperature T,

$$d = 9.72 \times 10^{-10} T^4 \ \mu\text{m}$$

Again putting $T = 500$ K as the upper limit for survival, we obtain $d = 60.8$ μm. Such a size permits the safe entry, not merely of individual bacteria but of whole colonies of bacteria, such as have been found, for example, in South Victoria Land by I.L. Uydes and W.V. Vishniac.[1]

Consider now the encounter of the Earth over a time interval of a few days or weeks with a cloud of cometary bacteria. Those entering at suitable glancing angles into the atmosphere, usually

[1] *Extreme Environments*, ed. M.R. Heinrich (London: Academic Press, 1975).

about 1% of all the particles encountered, will survive heating provided their sizes do not exceed the limits calculated above. They will then proceed, after being slowed, to fall gently to ground-level.

Figure 9.1 shows results calculated by F. Kasten for the fall *in still air* of spheres with a density equal to water.[1] The downward speed becomes less the lower the particles fall, and at first sight one might think that the main fraction of their times of fall to ground-level would be through the final 20 km. This is not so, however, because of air motions and because of water which condenses around the particles, thereby increasing their effective radii

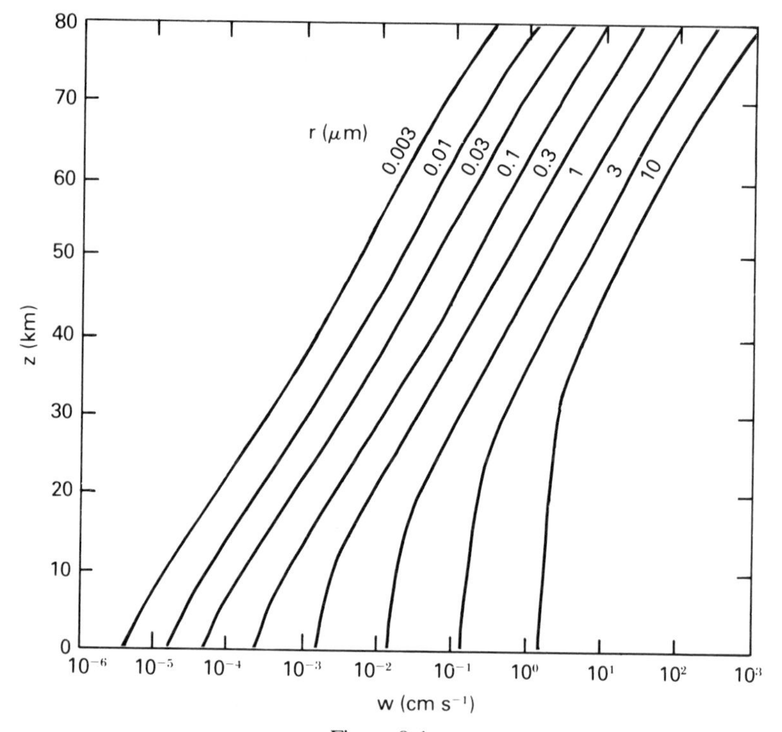

Figure 9.1
Falling speed w of spherical particles of various radii r as a function of height z.

[1] *J. Appl. Meteorol.*, (1968), 944.

considerably, and so causing them to fall more quickly. The slowest part of the journey to ground-level is over the height range from about 30 km down to 20 km.

Using a radius $r = 0.5\ \mu m$ for micrococci, and the same value of 0.5 μm as a typical average dimension for rod-shaped bacteria, the speed of fall at a height above ground-level of 20 km is seen from Fig. 9.1 to be about 10^{-2} cm^{-1}. Descent through an effective height interval of 10 km would therefore take about 10^8 s, i.e. about three years.

This is the length of time that would need to elapse before we could expect the atmosphere to have become cleared of its encounter with the cloud of cometary particles. Owing to air movements — the air is never exactly still as the calculations leading to Fig. 9.1 assumed — the particles would not all reach ground-level together, even though they all came into the atmosphere at more or less the same time. Arrival at ground-level would be spread over months, or even over a year or two, and would be spotty in a geographical sense, with some places receiving the particles before other places.

The details of arrival at ground-level evidently depend on the detailed meteorology of the atmosphere, and in particular on the extent to which major storms cause interchanges of air between the lower 10 km and the higher regions above 10 km. Such interchanges vary with the seasons in a more or less regular way, but superposed on the seasonal variations there are irregular effects, such as shifts of latitude in a particular year of the upper atmospheric jet streams. An extreme example of an irregular effect would be the explosion of a volcano, which can cause air interchanges to take place quickly even to heights above 30 km. The precise details of the incidence of a bacterium to ground-level must therefore be complex, both on a large continental scale and with respect to quite local variations, since in its final details the incidence must be dependent upon very fine-scale weather patterns, and indeed on the vagaries of all the winds that blow.

So far we have said nothing about viruses. The entry problem into the atmosphere is in principle less difficult for viruses than bacteria, just because viruses are smaller, and in general small particles experience less heating than larger ones. Whereas bacteria with sizes of about a micrometer are not heated to more than 500 K, particles a millimetre in size are heated to about 3000 K, and are then visible as shooting stars. This is an example of the effect of size on temperature, an effect clearly contained in the equation $d = 1.52 \times 10^{-15} T^4$ cm.

However, in order for viruses to survive irradiation by solar ultraviolet light it is necessary that they be encapsulated in a protective covering. This prevents them from gaining the full benefit of their smaller sizes, since heating on entry into the terrestrial atmosphere is then determined, not by the sizes of the viruses, but by that of their protective coverings. Indeed, in order to be able to take advantage of the enzymic repair mechanisms of bacteria it would be best for viruses to use larger cells as their protective coverings, in which case the entry problem into the atmosphere is just the same for viruses as it was in the above discussion for bacteria. Viruses will survive provided their hosts survive.

It is to be noted that although the terrestrial atmosphere is a hazard for entering bacteria and viruses, without it safe arrival at ground level would not be possible at all. Incoming small particles would impact the hard ground directly, and would be instantly gasified, as they must on striking the Moon. It is evidently useless to look for space-incident bacteria and viruses on the Moon or Mercury, although on Mars and Venus safe incidence must be possible. Whether subsequent survival on the Martian surface, or in the high Venusian atmosphere, is possible raises quite different and interesting problems that will be considered in Chapter 12..

Returning to the Earth, let us suppose that atmospheric eddies produce a patchy situation, shown schematically in Fig. 9.2, in which the dark areas are the regions of fall of a disease-causing

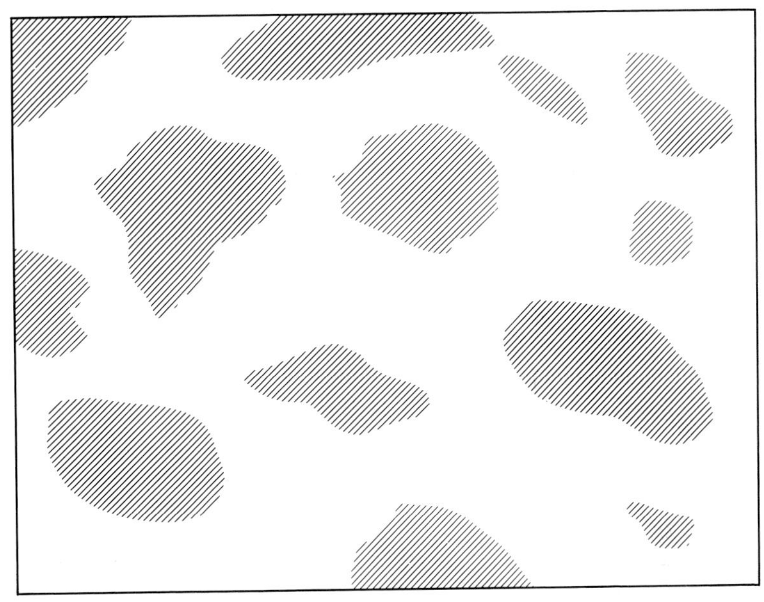

Figure 9.2
Schematic representation of pathogenic clouds settling at ground level. The patches (shaded areas) cover one third of the total area.

virus or bacterium. For a human disease, a person living a sedentary life may be engulfed by one of the dark spots of Fig. 9.2, or alternatively the place of residence may lie in a vacant area of the figure. There is a chance of contracting the disease and a chance of avoiding it, the chances being determined by the fraction of the ground area covered by the dark blobs in the figure. The persistent traveller, however, even if fortunate enough to be initially in a clear area, will sooner or later move into one or other of the pathogenic patches. Travellers — seamen, for example — are therefore more vulnerable to catching diseases than, say, farmers. In the days before air travel, when boat journeys with long periods at sea were frequent, there were often reports of epidemics occurring many days after leaving the last port of call. After recovering, passengers would sometimes to their astonishment find the disease had already arrived ahead of them at the next port of call.

Where a pathogen falls over an entire area, Fig. 9.2 is still likely to be representative of the situation at a particular moment, but with the detailed positioning of the dark blobs then changing with time. In this more extreme case, all who are in the area, traveller and sedentary person alike, are eventually hit by the bacterium or virus. The difference now is that, whereas the traveller is almost always hit early, the sedentary person may be hit either early or late, more likely later if the dark patches at any one moment form only a modest fraction of the total area. This relation of pattern-to-movement explains how it comes about that people in different countries, and even within different regions of the same country, are hit by diseases at times which are weeks, and sometimes months, apart. The conventional notion that one region 'infects' another is strained by the long time intervals that are often involved, contrasting with the great speed of modern transport. It takes half a day to fly from Hong Kong to London, but it took the so-called 'Red flu' virus of 1977 several months to make that same journey.

Among the animals, far-flying birds are particularly exposed to attack, since they are inevitably going to move sooner or later into one or more of the dark areas of Fig. 9.2. Hence we may expect birds to have evolved better immunities than more or less sedentary land animals have done. For influenza, the most carefully-studied cross-species disease, this is so.

The patches of Fig. 9.2 can be generated at ground-level by local obstacles — hills, woodland areas, buildings — as well as by natural eddies of the air. As another example, hot air near a furnace must tend to destroy pathogens, thus producing a clear area in the figure. A very fine-scale patchiness is therefore to be expected, especially over heavily-populated urban areas.

An incoming virus or bacterium may or may not be attacking; there are many non-pathogenic varieties of bacteria that simply lie around in the soil indefinitely, if not destroyed by lack of water or by the temperature of their surroundings. Even the attacking varieties could fail in some cases to interact with terrestrial life-forms, through being washed into rivers and thence into the sea. But where attack occurs, the bacterium or virus in question succeeds in establishing for itself a reservoir that grows through multiplication of the pathogen in the host plant or animal.

The reservoir which an attacking virus or bacterium succeeds in establishing for itself may be evanescent or it may be long-lived. The influenza virus in humans has not been found to persist for more than a few weeks, whereas the reservoir of the influenza virus in birds is longer-lived. Before the advent of modern health procedures, the smallpox reservoir persisted in humans on a time-scale certainly of many years and probably of some centuries. The reservoir of the virus *Herpes simplex*, which causes cold sores on the lips, is persistent on a still longer time-scale, probably of many thousands of years. Yet in our view every pathogenic reservoir eventually dies away if it is not renewed from outside, in a time interval much shorter than the times required for major evolutionary changes among the higher animals. We would argue that ten thousand years ago there were no reservoirs for most of the infectious diseases that afflict mankind today. In relation to our point of view on this question we describe in the next chapter what we have been able to uncover concerning the history of disease.

10

EVIDENCE FROM THE MEDICAL ANNALS

We began our enquiry into the history of diseases in a curious way. One of us was unable to recall any mention in Shakespeare of the common cold. The *Oxford Dictionary of Quotations* has 67 pages from Shakespeare, but nothing related to the common cold, although later authors with only a page or two in the *Dictionary* do make clear references to 'cold' as an affliction. Shakespeare's use of 'cold' always relates either to death or to the physical act of being cold. We could think of instances where 'fever' was used, but it is characteristic of the common cold that a clearly noticeable 'temperature' is comparatively rare.

Yet the common cold is not a new visitor to our planet, since there is the following description of it in the Hippocratic writings:

> 'In the first place, those of us who suffer from cold in the head, with discharge from the nostrils, generally find this discharge more acrid than that which previously formed there and daily passed from the nostrils; it makes the nose swell, and inflames it to an extremely fiery heat, as is shown if you put your hand upon it. And if the disease be present for an unusually long time, the part actually becomes ulcerated, although it is without flesh and hard. But in some way the heat of the nostril ceases, not when the discharge takes place and the inflammation is present, but when the running becomes thicker and less acrid, being matured and more mixed than it was before, then it is that the heat finally ceases.'

Evidently there were sufferers from the common cold in classical Greece, just as there are today. Nowadays more working days are lost from colds than from any other cause, a situation that possibly did not obtain in Shakespeare's time. It is true that Marion's trouble in the following passage from *Love's Labour's Lost* sounds rather like a cold, although there are other possible interpretations:

> 'While greasy Joan doth keel the pot.
> When all around the wind doth blow,
> And coughing drowns the parson's saw;

And birds sit brooding in the snow,
And Marion's nose looks red and raw,
When roasted crabs hiss in the bowl.'

Yet if indeed colds were then widespread, it is surprising that Shakespeare did not make use of them in his comic scenes.

The literary evidence for smallpox being a periodic visitor to our planet is stronger than for the common cold. A well-known medical text that we consulted at an early stage contained the remarkable statement:

> 'In the sixth century smallpox prevailed and subsequently, at the time of the Crusaders, became widespread.'

One could wonder how a disease as infectious as smallpox could have 'prevailed' in the sixth century and yet not have become widespread at that time. After telling the reader that the disease 'spreads like fire in dry grass', the text went on to say:

> 'The disease smoulders here and there and, when conditions are favourable, becomes epidemic.'

One can also wonder how it came about that something which spreads like fire in dry grass managed to smoulder.

The etymology of the Latin word *variola*, meaning pock-mark, is interesting. The purchaser of the 1854 edition of *Cassell's Latin Dictionary* in 896 pages would seem to have obtained excellent value for money, for as its author remarks in his preface:

> 'It [the dictionary] comprehends every word used by the following authors: Cato, Cicero, Caesar, Celsus, Columella, Catullus, Horace, Juvenal, Livy, Lucretius, Martial, Nepos, Ovid, Plautus, Phaedrus, the two Plinies, Persius, Propertius, Quintilian, Sallust, Seneca, Silius Italicus, Statius, Suetonius, Terence, Tibullus, Tacitus, Virgil, Vellejus, Varro, Vitruvius, Valerius Maximus, and Valerius Flaccus.'

None of these authors used the word *variola*, or any word for pock-mark. The inference is that smallpox did not exist in classical Rome.

If one presses the point by seeking the Latin for pock-mark in the English-to-Latin section of Cassell's dictionary, sure enough there is *variola*, but with an asterisk that is said to denote words of 'modern origin'. By modern origin is meant medieval Latin.

On this evidence we predicted that there would be no word for pock-mark in ancient Greek, and the classical scholars we have consulted assured us that this is indeed the case. Nor is there any description of smallpox in the Hippocratic writings. The Hippocratic descriptions were capable of being very clearly expressed, as the following shows for the case of mumps:

> '... Many people suffered from swellings near the ears, in some cases on one side only, in others both sides were involved. Usually there was no fever and the patient was not confined to bed. In a few cases there was a slight fever. In all cases the swellings subsided without harm and none suppurated as do swellings caused by other disorders. The swellings were soft, large and spread widely; they were unaccompanied by inflammation or pain and they disappeared leaving no trace. Boys, young men, and male adults in the prime of life were chiefly affected ...'

It defies commonsense to suppose that ancient Greek doctors would trouble themselves to observe mumps so carefully and yet would by-pass the deadly disease of smallpox, for which there is no cure or treatment, and which kills some forty per cent of its victims. Survivors are often so seriously disfigured that it also defies commonsense to suppose that classical authors would have made no mention of the disease, if it had then existed.

Lesions in the skin of mummies dating from 1550 to 1100 BC, spanning the time of Moses, have been interpreted as smallpox, although the existence of the disease is not confirmed in Egyptian papyri. Nor is there mention contemporaneously of it in India, while the situation in China is highly ambiguous at that date. Only some fifteen hundred years later does Ko Hung (*ca.* 330 AD) give a description of a disease that was probably smallpox.

A great plague broke out in Athens in 430 BC. The Pelopponesian War, which marked the decline of Athens, had begun a year before. The history of the war down to 411 BC was written by Thucydides with what has been called minute and scientific accuracy. He describes the plague, giving a detailed account of its symptoms. Some have identified the disease with smallpox, but if so there was no spread like fire in dry grass. Sparta would seem to have escaped its ravages, an advantage for the conduct of the war that some authorities consider to have been decisive.

The precision of Thucydides has provided a challenge of diagnosis to doctors down the ages. After reviewing various diagnoses, one commentator remarked:

> 'I have looked into many professional accounts of this famous plague, and writers, almost without exception, praise Thucydides' accuracy and precision, and yet differ most strongly in the conclusions they draw from the words. Physicians — English, French, German — after examining the symptoms, have decided it was each of the following: typhus, scarlet, putrid, yellow, camp, hospital, jail fever; scarlatina maligna; the Black Death; erysipelas; smallpox; the oriental plague; some wholly extinct form of disease. Each succeeding writer at least throws doubt on his predecessor's diagnosis.'

In a recent book[1] we accepted this confusion over diagnosis as showing the Plague of Athens to have been a disease unknown in modern times. Now we are inclined to take the suggested association with smallpox rather more seriously, however. What Thucydides wrote[2] was:

[1] *Diseases from Space* (London: Dent, 1979).
[2] Translated by Ben Jowett.

'The season was universally admitted to have been remarkably free from other sicknesses; and if anybody was already ill of any other disease, it finally turned into this. The other victims who were in perfect health, all in a moment and without any exciting cause, were seized first with violent heats in the head and with redness and burning of the eyes. Internally, the throat and the tongue at once became blood-red, and the breath abnormal and fetid. Sneezing and hoarseness followed; in a short time the disorder, accompanied by a violent cough, reached the chest. And whenever it settled in the heart, it upset it; and there were all the vomits of bile to which physicians have ever given names, and they were accompanied by great distress. An ineffectual retching, producing violent convulsions, attacked most of the sufferers; some as soon as the previous symptoms had abated, others, not until long afterwards. The body externally was not so very hot to the touch, not yellowish but flushed and livid and breaking out in blisters and ulcers. But the internal fever was intense; the sufferers could not bear to have on them even the lightest linen garment; they insisted on being naked, and there was nothing which they longed for more eagerly than to throw themselves into cold water; many of those who had no one to look after them actually plunged into the cisterns. They were tormented by unceasing thirst, which was not in the least assuaged whether they drank much or little. They could find no way of resting, and sleeplessness attacked them throughout. While the disease was at its height, the body, instead of wasting away, held out amid these sufferings unexpectedly. Thus, most died on the seventh or ninth day of internal fever, though their strength was not exhausted; or if they survived, then the disease descended into the bowels and there produced violent lesions, at the same time diarrhoea set in which was uniformly fluid, and at a later stage caused exhaustion, and this finally carried them off with a few exceptions. For the disorder which had originally settled in the head passed gradually through the whole body and, if a person got over the worst, would often seize the extremities and leave its mark, attacking the privy parts, fingers and toes; and many escaped with the loss of these, some with the loss of their eyes. Some again had no sooner recovered than they were seized with a total loss of memory and knew neither themselves nor their friends.'

Comparison with a modern medical text shows that a tolerable correspondence exists between most of the positive assertions of Thucydides and the characteristics of the severest form of smallpox, the confluent form. Headache, high temperature, smell, hoarseness, ineffectual retching, and thirst are all found in confluent smallpox. The words 'breaking out in blisters and ulcers' implies pustules in the skin rather than a rash or spots on the skin, and this too is correct. Most diagnostic is the passage: 'While the disease was at its height... finally carried them off with few exceptions.' Thus in a modern text:

> 'In fatal cases by the tenth or eleventh day the pulse gets feebler and more rapid, the delirium is marked, there is sometimes diarrhoea, and with these symptoms the patient dies.'

It is also correct that eyes could be 'lost' through subsequent blindness and that fingers and toes could be lost through gangrene, although these complications are apparently not common in modern times. Post-febrile insanity is sometimes met, which could conceivably explain the last sentence of the quotation: 'Some again had no sooner recovered ...'

Correct further statements appear in later paragraphs of Thucydides. Thus:

> 'Equally appalling was the fact that men died like sheep, catching the infection if they attended on one another; and this was the principal cause of morality.' ...
>
> 'For no one was ever attacked a second time, or not with fatal result.'

Omissions rather than positive assertions give the main reason for doubting the association with smallpox. There is no mention of the severe lumbar pains which occur at an early stage of the disease, contradicting the sentence: 'For the disorder which had originally settled in the head...' More important, there is no mention of the very high density of pustules on the face that is the main outward visual characteristic of confluent smallpox. Nor is there mention of the scarring and disfigurement of survivors. It hardly seems credible that an accurate observer would have by-passed the most terrible aspects of confluent smallpox with the single remark of the 'body breaking out in blisters and ulcers'.

In these circumstances it seemed desirable to return to the original Greek of Thucydides, on the interpretation of which we are indebted to Dr Humphrey Palmer of University College, Cardiff. We understand that the Greek carries the implication of blisters with an initial appearance similar to those sometimes experienced by oarsmen, and that the word used for 'ulceration' would be consistent with the subsequent development of pustules. This is more informative than the English translation, and would accord better with the eruption of smallpox.

The phrase rendered by Ben Jowett as 'most died on the seventh or ninth day of internal fever' has always caused difficulty for translators. Faced with the improbability that death somehow skipped the eighth day, some have taken the liberty of changing nine to eight, 'some died on the seventh or eighth day', thus presuming that the father of history was unable to count! What Thucydides actually wrote was 'most died in the ninth *and* seventh days ...' We understand from Dr Palmer that a controversy existed in the late fifth century BC as to whether weeks should be counted in blocks of seven days or nine days. Thucydides may therefore have meant 'most people died in a week plus a week', and that by using two different lengths for the week he intended his statement to be approximate to within about two days. Deaths from confluent smallpox do in fact mostly occur after roughly two weeks. Less frequently, there are deaths after about five days, and these are from heart failure, which is just what Thucydides says: 'whenever it settled in the heart, it upset it'.

If the Plague of Athens was indeed confluent smallpox, several interesting points follow. We really could then conclude that smallpox was otherwise absent from ancient Greece and Rome, since Thucydides and other classical authors write about the Plague of 430 AD as a unique event. Second, it did not spread into the Peloponnese, and so was not like fire in dry grass. Third, it did not maintain itself. Livy writes of it as lasting for four years only.

The Antonine pestilence which spread in the second century AD through Italy and Western Europe has been ascribed to smallpox on the strength of a description by Galen. If the disease was smallpox then, it would seem to have died out in spite of its infectious quality, for there is no mention of it by other writers of the early Christian era (Celsus, Aretaeus).

Smallpox appeared unequivocally in Arabia during the sixth and seventh centuries. The Caleph Yezid who died in 683 AD was said to be pitted because of it. Epidemics described by Gregory of Tours in the sixth century and by the Venerable Bede in the seventh may also have been smallpox. The disease is caused by a virus that has no known host except man, which circumstance, taken with the nature of the disease itself and with the historic evidence, sets severe problems. To maintain itself, the virus (according to the usual point of view) had to be always passing to new victims as the old victims either died or recovered, since a person who has recovered does not continue to harbour the virus. Thus the virus would become extinct should the chain of case-to-case transmission ever be interrupted. The virus would also become extinct in a situation where on the average each victim infected fewer than one other person, for then the number of contemporaneous victims would decrease steadily to zero along the transmission chain. In the opposite situation, where on the average each victim infected more than one other person, the number of contemporaneous victims would grow along the transmission chain. There would be an epidemic, increasing until the virus at last exhausted the supply of susceptible people.

The condition for smallpox to smoulder over long periods of time is not easy to achieve. It requires a precisely-controlled state in which each victim infects just one other person, an unstable condition for a highly infectious disease, except perhaps in the following special situation. Imagine the disease to have run through nearly all the population of some city. The survivors have become immune to the virus and they do not harbour it, so that only the remaining low density of susceptibles is involved. If it were not for births constantly tending to increase the density of susceptibles, the disease would die away. But new births maintain the supply of victims so that the disease propagates itself at a steady level. The disease is then endemic in the city, which becomes a 'focus' from which it can spread to other uninfected populations.

A city endemic to smallpox would be exceedingly noticeable, however, since a large fraction of its population would be survivors disfigured by the disease. There could have been no such city known to Greek or Roman writers, otherwise its exceedingly repugnant nature would surely have been remarked upon. Nor does it seem as if there could have been such a city in India or China, otherwise writers there could hardly have failed to leave a record of it. What is needed is a hidden city, a mystery city, as unknown as a distant comet in space.

Until some 6000 years ago there were no cities. Over much of the greater part of our history the density of the human population was so low that the smallpox virus could never have found a steady supply of victims. Nor could most of the diseases which afflict us today. Whence then have they come? There is, within the usual point of view, no plausible answer to this question. The usual apology of an answer is to suppose that virulent human pathogens must have had some other means of survival in prehistoric times than they have today.

It is popularly supposed that viruses, bacteria, and protozoa are capable of evolving almost instantly to fill any environmental niche that may present itself, and if this were true in a fundamental genetic sense the situation for the usual point of view might not be so bad. But in spite of a veritable torrent of experiments in the laboratory, such changes have not been demonstrated. Changes in the laboratory are of a negative rather than a positive character, as for instance *Drosophila* (fruit fly) was changed by squirting it with X-rays, thus tearing up its genetic material. Changes of an apparent kind could also be generated in a microbial population as a result of selective pressures impressed on it by the environment. Thus the environment could, under appropriate conditions, sieve out even trace components of an initial genetic mix, giving a semblance of rapid innovative change.

The same is also true for viruses. Under natural conditions pathogenic viruses are supposed to undergo all manner of weird and wonderful changes. Yet a genetically well-defined virus injected into a laboratory animal comes out (after multiplication in the animal) as essentially the same virus that went in. Laboratory workers are at their wits end to understand the bewildering changes of viruses like those of influenza and the common cold, changes that seem to be taking place with the greatest of ease in the outside world. Nothing ever happens in the laboratory itself, yet pandemonium seems to occur in the local cafeteria and in the home. Like a well-known kind of dream, when you look nothing moves, but whenever you look away for the briefest moment everything changes.

The situation is easily explicable if one supposes that viruses, bacteria, and protozoa are continuously incident on the Earth from space, with a range of biochemical properties significantly wider than the habitats available on the Earth can admit. The minority that succeed in fitting themselves to terrestrial habitats fill all the available niches. Should a new habitat arise, through men beginning to dig coal out of the ground for example, then immediately a life-form already present among the infalling types fills the new niche. The process is essentially instantaneous, and it needs no terrestrial evolution at all.

For pathogens, the pattern is one of never-ceasing change, due in part to the varying immunities of their hosts and in part to an ever-changing distribution of the pathogens themselves. Human populations were exposed to smallpox in the sixth and seventh centuries because it was then that the smallpox virus was incident from space. Classical times were free of the disease because the Earth did not happen to encounter the virus over a millennium from about 800 BC to 200 AD.

A few pathogens are rare visitors to our planet, a possible example being the one responsible for the Plague of Athens. Another was the bacterium or virus responsible for the so-called 'English Sweats', which came in five epidemics between 1485 and 1552. The symptoms, besides sweating, were thirst, nausea, and fever. Unlike smallpox, an attack of the disease did not confer immunity against subsequent attacks. Many thousands are said to have experienced repeated bouts, and perhaps a million people are thought to have died from the disease, which, although apparently first described in London, was present also in Western Europe. The 'English Sweats' disappeared as suddenly as they had appeared, as if miraculously prohibited.

Man is the only known host of the measles virus, and as with smallpox a single attack confers a lifelong immunity. Measles is also a disease in which the virus disappears when recovery or death occurs. Consequently much the same problems exist for measles as for smallpox, particularly as the histories of the two diseases are remarkably similar. There is no description of measles either in Hippocrates or in the Indian medical writings of the sixth century BC. Since the rash associated with measles is very characteristic, this surely implies absence from ancient Greece and Rome. There is uncertain evidence of its appearance in the second century AD, and more certain evidence from the sixth century. As with smallpox, measles was clearly described by Abu el Rhazes (860–932 AD). Rhazes seems to have considered the two diseases of comparable severity, perhaps indicating that measles was then attacking adults who had not acquired immunity in childhood.

The World Health Organization has recently declared the world to be free from smallpox, as a result of quarantine measures and of an intensive vaccination programme. One can wonder why the world has not similarly been made free from measles. The incubation period for measles is even longer than for smallpox, about 14 days, so that quarantine measures for measles should be similarly effective. An effective vaccine for measles also exists. Faced with the contrast between the eradication of smallpox and the ever-present quality of measles, we would argue that the measles virus is continually incident from space, whereas the smallpox virus was a survivor from incidence in the past. According to the conventional point of view no explanation can be offered, except perhaps to say that the world is too indifferent towards measles for a serious attempt at its eradication to be made. We doubt that this explanation is correct.

Turning now from virus-induced diseases to those caused by bacteria, the most famous example is that of the Black Death, or bubonic plague. Unlike the cases considered above bubonic plague is not primarily a disease of man. The bacterium *Pasteurella pestis* attacks many species of rodent, its first target. Because the black rat happened to live in close proximity to humans, nesting in the walls of houses, the physical separation of people from the black rat was small and could be bridged by fleas, which carried the bacterium from the blood of the rats to the blood of humans. This transfer process was not welcomed by the fleas, who preferred to stay with the rats, quitting them only as their hosts died from the disease.

Although human–flea–human transfer of the bacterium presumably occurred, it does not seem to have been sufficient to maintain the disease, which died out as the supply of rats became exhausted. The saving grace was certainly not the prayers that were said everywhere in the churches. By an accident of providence the fleas did not fare well on a diet of human blood. The affliction was irrationally thought to be a punishment from God imposed (like the Flood) as a reprisal for human wickedness. Yet it was precisely in the institutions of the Church where the death rate was highest, particularly among monks in the monasteries, crowding together for their prayers.

As with smallpox, bubonic plague has come in sudden bursts separated by many centuries, and there are the same difficulties of understanding where *Pasteurella pestis* went into hiding during the long intermissions. A somewhat ambiguous reference occurs in the Old Testament, at a date of about 1200 BC, when the Philistines are said to have been attacked by 'emrods [buboes] in their secret parts ...' as a reprisal of God for an attack on the Hebrews. A clear reference to plague occurs in an Indian medical treatise written in the fifth century BC, in which people are advised to leave houses and other buildings 'when rats fall from the roofs above, jump about and die'.

There may have been an outbreak of plague during the first century AD, with centres of the disease in Syria and North Africa, but between the first and sixth centuries there were no known attacks of the disease. In 540 AD, a pandemic involving the Near East, North Africa and Southern Europe is said to have had a death toll that reached 100 million, with more than 5000 dying each day in Constantinople alone. This was the so-called Plague of Justinian, the Roman Emperor of the time.

Bubonic plague would then seem to have disappeared from our planet for eight centuries, until it reappeared with shattering personal and social consequences in the Black Death of 1347–50. Thereafter, the disease smouldered with minor outbreaks until the mid-seventeenth century when for two centuries it seemed once again to have died out, only for it to reappear in China in 1894. In India, it killed some thirteen million people in the years up to the first world war.

By working from historical records of the first outbreaks of plague in various population centres, Dr E. Carpentier obtained contours (Fig. 10.1) showing the spread of the Black Death across Europe, except that the one for 31 December 1347 appears to have

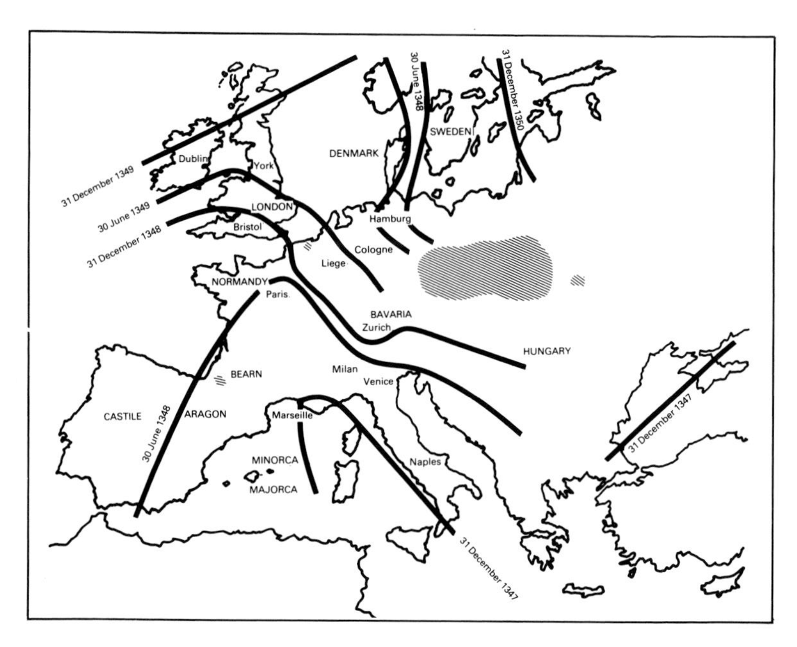

Figure 10.1
The spread of the Black Death in Europe (after Dr. E. Carpentier).
The contour lines indicate the broad spread of the plague.

been drawn to fit a belief rather than from documentary facts. The belief is that the disease was brought into Europe from the Tartars who had besieged the Genoese base of Caffa in the Crimea. The line of December 1347 is said to mark the voyage of Genoese ships back to Italy. One can have some doubt about it, however, since the much better attested contour for December 1348 seems to be headed to intersect the one for December 1347 somewhere in the region of the Danube delta, which from the nature of the contours — the *first* outbreaks at the various locations — is an impossible condition.

The contours of Fig. 10.1 are interpreted by orthodox opinion as steps in the march of an army of plague-infested rats. Humans with the disease collapsed on the spot, and we think afflicted rats must surely have done the same. To argue that stricken rats set out on a journey that took them in six months, not merely from southern to northern France, but even across the Alpine massif, borders on the ridiculous. Nor does the evidence by any means support in detail such an inexorable stepwise advance of the plague. Pedro Carbonell was the archivist to the Court of Aragon, a post that in its very nature could only be held by a person with a clear appreciation of the difference between fact and fiction. Carbonell reports that the Black Death began in Aragon, not at the Mediterranean coast or at the eastern frontier, but in the western inland city of Teruel.

Since it apparently stretches credulity too far to argue that the advancing army of stricken rats also managed to swim the English Channel, it is said that the Black Death reached England by ship. Yet the contours of Fig. 10.1 are of quite the wrong shape for boats to have played a significant role in spreading the disease. If *Pasteurella pestis* had been carried by sea, the earliest contour would be wrapped around the coastline from the Mediterranean to northern Europe, with subsequent contours then filling in gradually towards central Europe. Not only this, but if the bacillus had travelled by sea, the coast of Portugal would have been seriously affected, whereas the evidence is that the plague scarcely penetrated to Castile, Galicia, and Portugal.

There are many descriptions of communities that isolated themselves deliberately from the outside world, many such descriptions from English villages. Yet isolation was to no avail. The Black Death would strike suddenly, and within a week the people in such a community would be just as affected by the disease as everyone else.

What remarkable rats they were. To have crossed the sea and to have reached into remote English villages, and yet to have effectively by-passed the cities of Milan, Liège, and Nuremberg! To have reached into remote villages and yet to have largely spared the shaded areas of Fig. 10.1, especially the extended area in

Bohemia and southern Poland. The astonishing reason offered for this behaviour is indicative of the state of mind engendered by orthodox theory. The rats, it is said, disliked the food available to them in these regions.

We remarked above that Indian doctors had noticed already in the fifth century BC the connection of plague with rats. Yet medieval doctors had no such thoughts. It was their overwhelming view that the pestilence had its origin in the air — 'poisoned' air was the widely favoured explanation. It has been fashionable to decry this view as an unsubstantiated superstition, although the state of technical understanding in fourteenth century Europe was higher than it had been at any earlier stage of human history. Indeed, unless one is prejudiced by modern superstitions, the contours of Fig. 10.1 are a clear indication that *Pasteurella pestis* hit Europe from the air. There was no marching army of plague-stricken rats. The rats died in the places where they were infected, just as humans did. By falling from the air, *Pasteurella pestis* had no difficulty at all in crossing the Alps, or in crossing the English Channel. Remote English villages were hit, however determinedly they sought to seal themselves off from the outside world, because the plague bacillus descended upon them from above; and against an aerial assault all the precautions taken were of no consequence. Milan, Liège, and Nuremberg went comparatively unscathed because it is in the nature of incidence from the upper atmosphere that there will be odd spots where a pathogen does not fall, as shown in Fig. 9.2. So too did Bohemia and southern Poland escape, even though these areas grow food just as palatable to rats as everywhere else.

We conclude this chapter with a few remarks about Whooping Cough or Pertussis. This is a bacillus-caused childhood disease that has been well recognized since 1578, but which probably existed much earlier. It has been known for a while that epidemics of whooping cough occur in approximately 4-year cycles. The conventional wisdom has been that this cyclic behaviour is due to the lag time required for the build-up of a new generation of susceptible children following an epidemic of the disease. An epidemic outbreak is thought to reduce the population density of susceptibles to such a low level that transmission through person-to-person contact is effectively stopped. The disease then smoulders until the susceptible population density builds up again from new births and migrations to a critical value — presumably every four years.

An interesting situation arose for testing this theory because the population density of children susceptible to whooping cough was controlled during the period 1960–1975 using a vaccine that conferred active immunity against the disease. During these years the susceptible population fell essentially to zero. If the immunization was 100% no epidemics should have occurred. More realistically, in the practical situation where the immunization was not 100% across the population, epidemics should according to the conventional theory now have been much more widely spaced in time than in the pre-immunization era. As we can see in Fig. 10.2 the 4-year cycle continued unabated throughout, except that fewer children were involved in each epidemic. We note also that the decline of popularity of the whooping cough vaccine (because of

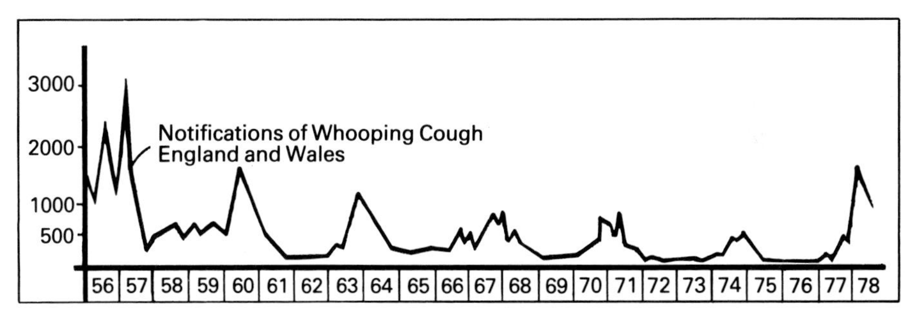

Figure 10.2
The incidence of whooping cough in England and Wales during the period 1955–1978. (Courtesy of *The Sunday Times*)

alleged complications) which occurred in the late 1970s has done nothing to alter the nearly clock-work periodicity of whooping cough extending over several periods. The numbers involved in the 1978 epidemic increased again, but the period of separation between epidemics was the same. The approximately constant periodicity of whooping cough epidemics under conditions of grossly varying population susceptibility would point to the causative bacillus being a recent space traveller.

11
INFLUENZA

REPORTS SHOW EPIDEMIC IS MAKING GAINS

Hospitals Are Filled With Patients and Tents Have Been Ordered

MANY SCHOOLS CLOSE

New cases of influenza reported to the Board of Health during the 24 hours ending at 4 o'clock yesterday afternoon indicate that the epidemic has taken a turn for the worse in Pittsburg. For that period of time 813 new cases are on record, compared with 873 for 32 hours beginning Saturday noon. The total number of influenza cases was 4,448, while the new pneumonia cases number 40, making a total of 319. Deaths from pneumonia were 39, from influenza-pneumonia, 20, and from influenza, 5, making a total number of 74 for the 24

MAYOR ASKS COUNCIL FOR $100,000 TO HELP CHECK INFLUENZA

POST

73 Deaths and 1,200 New Cases Are Reported.

DISEASE MAY BE ON WANE

An emergency ordinance providing $100,000 with which to meet the demands of influenza was introduced in council yesterday by Mayor E. V. Babcock and Controller E. S. Morrow. The measure will be acted on today when, it is probable, the amount will be reduced in committee. Mayor Babcock said there was no immediate change in the situation and that the action was taken in line with the administration's ... proper preventative and ...

COUNTY TO GIVE AID

Annex to Courthouse Is Made Emergency Influenza Hospital

Pgh. Post October 26, 1918

Courthouse annex on Ross street will be opened this morning by the city as another emergency influenza hospital ... converted into an experimental laboratory

URGENT CALLS SENT OUT BY RED CROSS

More Men and Women Needed to Aid in Combatting Spanish Influenza.

VOLUNTEERING SLOW

CATHOLIC AID IN EPIDEMIC IS ACCEPTED

GAZETTE

Bishop Canevin Informs Health Director of 27 New Relief Stations.

SISTERS AS NURSES

... Remains, Babcock Goes To Harrisburg

GAZETTE

State Commissioner Refuses to Modify Health Regulations Here.

OCT 29 1918

FEW CASES FEWER

Encouragement Found in Reports on Influenza and

DRUG TRADE HAS BROKEN DOWN UNDER DEMAND FOR GRIP CURES

POST

One Wholesale House Compelled to Close Its Doors.

KINGSLEY HOUSE GETS PATIENTS

Pittsburgh's drug trade has all but broken down under the demands

WORST IS OVER IN EPIDEMIC, AVERS HEALTH OFFICIAL

CHRONICLE TELEGRAPH

New Cases On the Wane, But Deaths to Increase, Opinion of Dr. Philip E. Marks.

TOTAL OF PATIENTS, 14...

The worst is over in Pittsburgh's influenza epidemic, so far as ... are concerned, in the opinion of Dr. Philip E. Marks, head of the ... of infectious diseases, who ... that the situation so far ... much better than it was at the ... time last week. ... were reported today ...

INFLUENZA ORPHANS IN NEED OF HOMES

CHRONICLE TELEGRAPH

NORTHERN COUNTIES FOR AID TO CHECK GRIP EPIDEMIC

Pgh. Post

Pittsburgh Conditions Are Satisfactory, ... Officials Declare.

October 18, 1918

CHURCH SERVICES BANNED TODAY

Influenza conditions in Pittsburgh yesterday were such ... medical authorities were devoting much of their energies to preventing an epidemic outbreak in other parts of Western Pennsylvania. Pittsburgh has been ... to handle matters in the ... of the state, and calls were ... in from northern counties for assistance.

Dr. Adolph Koenig, county ... supervisor, is arranging to

CITY HEALTH BOARD MAY GET POWER TO LIFT BAN HERE SOON

POST

Epidemic Peak Reached Week Ago, Claim.

Influenza, as an epidemic, reached its peak in Pittsburgh on Tuesday, Wednesday and Thursday of last week, was the opinion of Dr. ...mer R. Batt, state registrar, ... an examination of the situation yesterday, and it is believed that ... report to Harrisburg will be followed by a decision to allow Pittsburgh authorities to dispose of the ban here at their pleasure.

There is no recourse within the orthodox view of things but to suppose that virulent viruses were carried through the scores of millennia of prehistory by animals, particularly by rodents and birds. Where then, one can ask, are the birds and rodents now attacked by measles and smallpox, and indeed by all other diseases which are specific to man? To this pertinent question it is answered that viruses change quickly, and that after adaptation to a new host they soon become ineffective against an old host. There is neither experimental evidence under controlled laboratory conditions, nor theoretical support for the supposition that particular strains of virus change quickly in their basic genetic properties, and there is a good deal of evidence to suggest that they scarcely change at all.

The generalized statement that viruses change quickly, because the viruses of influenza and the common cold are observed in nature to change quickly, is not the relevant issue. Viruses do not change quickly when they multiply under laboratory conditions. Influenza can be induced in humans or in experimental animals by inoculation with particular strains of virus. In such cases the virus which emerges after multiplication is essentially the same as the one which went in.

Viruses can take on non-genetic properties of their hosts, and such properties can of course change quickly, simply through a change of one host for another. An example of such a property is the outer layer of fatty material (lipids) which viruses acquire on emergence from the host. Detailed experiments with a virus (VSV) that causes an influenza-like disease in farm animals has shown that the virus wraps itself in lipids from the host.[1] The same is probably true for the actual influenza virus. Such changes are not of a basic genetic type, however, and it is genetic changes with which we are concerned here.

Yet the influenza virus does show enormous genetic variety. Isolations of the virus from different victims change markedly over

[1] 'The assembly of cell membranes', in H.F. Lodish and J.E. Rothman, *Scientific American*, 240 (January 1979), 38–53.

the years, and even contemporaneous isolations over small areas show variations that imply of the order of five hundred differences in the genetic sequencing of the virus. A simple calculation, using well-known estimates for the mutational rate induced by the copying of genetic material, shows that no normal multiplication of the virus could produce remotely so many differences. Indeed any virus that changed by hundreds of genetic bases in moving from one victim to another would almost surely become extinct through its own instability.

What the many varieties of the influenza virus clearly suggest is that this virus contains at least a component of it that has been recently incident from space. Outside the Earth, living cells and viruses or fragments of viruses which they may contain can be protected against solar ultraviolet light by travelling in clumps. These clumps of cells would be scattered into their individual members on entry into the Earth's atmosphere. The scattering would occur at an altitude of about 130 km, high above the ozone layer. From that altitude down to the protective ozone layer the fall time ranges from hours to several days, according to the size of the cell, and during this time exposure to ultraviolet light must usually take place. (There could be exceptions for large cells falling quickly on the night side of the Earth, and for cells falling in winter on to the polar areas). Ultraviolet damage must therefore often take place, both for cells and for any viral particles that may be riding with them. In some cases there will be enzymic repair, and then genetic variations must arise. One can expect almost every viable form of repair to be represented, and the number of such forms could be very large. As techniques for determining precise base-sequencing of genetic material improve we expect the class of influenza viruses to grow larger and larger, to a point where every influenza victim could turn out to have been infected by a form of the virus that is unique in its finest detail.

Unlike a laboratory experiment in which an animal is inoculated with a particular strain of virus, a natural attack involves a broad spectrum of varieties, under conditions that are well-suited to the occurrence of what are called 'recombinant' events. The genetic material of influenza comes in eight, separate, small chromosomes. Should two markedly different viral particles enter the same host cell, the chromosomes of the two varieties can mix, with the effect that a new strain can arise, with some chromosomes coming from one variety and the rest of the complement of eight coming from the other variety. A similar mixing could also occur between fragments of corresponding chromosomes in the two viruses, to produce hybrids at a more subtle level of recognition.

Major new strains of the influenza virus are known to have been produced in this way. Careful experimental work has shown *how* the mixing has taken place, but not *where* it took place. The popular view holds that recombinant events leading to major new strains of the human virus occur initially in other animals. Pigs and birds are the animals usually discussed. For reasons discussed later in this chapter, we doubt the correctness of this presumption. Our view is that the recombinant events take place on a large scale within humans themselves.

Consistent with its broad spectrum of varieties, the influenza virus attacks a wide range of animals — many mammals, birds, and possibly some reptiles. Each kind of animal acts as a selective agent, picking out a subset of varieties which attack the particular animal in question. These subsets mostly do not overlap each other, although cases of the same influenza variety attacking two different kinds of animal have occasionally been reported.

The selection imposed by an animal on the primary spectrum of influenza varieties operates in a complex way, partly through the detailed properties of the cells of the animal in question (e.g. through the cell wall and membrane) and partly through constantly varying immunological responses to the different varieties of the virus. Because of this complexity, the output of virus from an infected animal gives only a fragmentary picture of the original input. Indeed, whenever a recombinant event takes place within an animal, the output becomes different from anything present in the primary input. It is therefore an error of the first magnitude to suppose that input and output are the same, an error which seems to us to lie at the root of criticisms which have sometimes been made of our point of view. (A more detailed discussion of the relationship between input and output for influenza is given in a technical appendix at the end of this book.)

Although less information is available for the common cold than for influenza, it is well-known that there are also many varieties of the common cold virus, and this too is an indication that the common cold, or at least a considerable component of it, is also a present-day space traveller. Indeed all those diseases that vary in their intensities year by year, being sometimes mild and sometimes severe in their attack, are likely to be current space travellers. There is also a rough and ready rule, that the more difficulty the medical profession finds in coping with a disease, the more likely it is to be a current space traveller. Diseases that built reservoirs in the distant past will usually have become much more genetically uniform, since their initially large gene pool must tend to be selected with the passage of time into those strains which best attack their hosts. Partly because of their trend towards genetic uniformity, and partly because they are no longer being replenished, old reservoirs are exposed to extinction, and it is in such cases that the medical profession has had its greatest successes.

It is well-known that people differ, one to another, in the precise details of how they are attacked by coughs, colds, influenza, and gastric upsets. 'Things go round', hitting each of us a little differently from our neighbours. According to the usual lore such variations are supposed to be subjective, arising entirely from differences of susceptibility within ourselves. Where otherwise healthy people are involved, we suspect this lore to be wrong. Variations of our symptoms are caused, not so much by our subjective susceptibilities, as by genuine differences in the viruses by which we are severally attacked. In their fine genetic detail the viruses themselves are peculiar to each of us.

As well as having variations in their fine genetic detail, influenza viruses have larger differences which are classified into types A, B, and C, and into sub-types of these main classes. Sub-types of A are particularly common, and it was a new sub-type of A that first appeared in Sardinia in 1948. Commenting on this first appearance, Professor F. Magrassi wrote:[1]

> 'We were able to verify ... the appearance of influenza in shepherds who were living for a long time alone, in solitary open country far from any inhabited centre; this occurred absolutely contemporaneously with the appearance of influenza in the nearest inhabited centres'.

This observation shows that influenza can be contracted without connection being necessary to another human. To explain Professor Magrassi's finding we must assume one of the following to have been true:

(a) The virus responsible for the Sardinia outbreak came from space. It happened for meteorological reasons (perhaps heat from the nearby active volcano of Stromboli) to touch down first in the area of the Tyrrhenian Sea.
(b) Winds blew the virus from some other area on the ground.
(c) Some animal exuded the virus all over Sardinia, and it passed to humans all in the same moment, whether in the remote countryside or in populated centres.

Of these, (b) is implausible because the new influenza sub-type started in Sardinia. There had been no other outbreak on the ground from which it could blow. As regards (c), there have been few reported cases of humans contracting influenza from animals. Influenza is said to have passed from pig to man at Fort Dix, New Jersey, in 1976. We mention this incident for those who wish to believe it. At all events, observed transfers from animal to man are

[1] *Minerva med. Torino,* 40 (1949), 565.

so rare as to make it very doubtful that very many such transfers could have occurred in Sardinia, and all at effectively the same moment.

Birds rather than pigs would be the least improbable animal to have produced such transfers. One might suppose that flocks of birds arrived in Sardinia from somewhere harbouring the new sub-type of the virus, and that by broadcasting bird muck all over the island they contrived to produce the remarkable contemporaneous outbreak investigated by Professor Magrassi.

In 1918 there were only some 45 000 people living in Alaska, which is two-and-a-quarter times larger than the State of Texas. In November and December of 1918 a lethal epidemic of influenza passed over the whole of that vast thinly populated territory, when human travel from the coast to the interior was essentially impossible because of snow and ice. Here again then we have an example of the spread of influenza by some means other than person-to-person contact, and as before there are the same three possibilities for explaining how the spread occurred.

In this case, it is possibility (c) that one must reject. Flocks of birds did not arrive in frozen November and December to spread ubiquitous muck all over the huge territory of Alaska. But the winter jet stream could quite well have overturned the upper atmospheric air, causing a cloud of virus-bearing particles to descend on even so large an area as Alaska. Indeed it could have been the descent of bitterly cold air from the high atmosphere which caused Governor Riggs of Alaska to declare as follows to the Senate Committee of Appropriations (16 January 1919):

> 'You have the short days, the hard cold weather, and you only make 20 to 30 miles a day over the unbroken trails. The conditions there are such as have never happened before in the history of the Territory.'

The influenza pandemic of 1918 had other peculiarities. Dr. Louis Weinstein writes:[1]

[1] *New England Journal of Medicine*, (May, 1976).

'The influenza pandemic of 1918 occurred in three waves. The first appeared in the winter and spring of 1917–18... This wave was characterised by high attack rates (50 per cent of the world's population were affected) but by very low fatality rates... The lethal second wave, which started at Fort Devens in Ayer, Massachusetts, on September 12, 1918, involved almost the entire world over a very short time... Its epidemiologic behaviour was most unusual. Although person-to-person spread occurred in local areas, the disease appeared on the same day in widely separated parts of the world on the one hand, but, on the other, took days to weeks to spread relatively short distances. It was detected in Boston and Bombay on the same day, but took three weeks before it reached New York City, despite the fact that there was considerable travel between the two cities.'

Here we have a third case where influenza was not spread by person-to-person contact, for nobody in 1918 could travel from Massachusetts to Bombay in the day or two which elapsed between the appearance of the new wave at Fort Devens and its appearance in Bombay. Nor could the fastest flying birds, the shearwater, swift, or even the albatross, make that journey in the time, even if Boston and Bombay were on the customary flight paths of these birds which they are not. Nor are there winds that blow over such a route, with such a speed, and over such a distance.

A markedly new variety of influenza falling from the high atmosphere will in general arrive at ground-level at different places at different times. There will be a place where it arrives first, and this will be where the new epidemic begins, as in September 1918 at Ayer, Massachusetts. That the virus should happen to touch down shortly afterwards at two such widely separated places as Boston and Bombay does not strain credulity at all. It was just that two of the first major patches of a virus which eventually descended on the whole world happened to hit Massachusetts on the one hand and the Bombay area on the other. Nothing travelled horizontally between these two afflicted regions.

Is influenza ever spread by person-to-person contact? Although medical opinion holds emphatically that it is, nothing that we have found in the literature proves that this opinion is correct. Doctors are familiar with sudden outbreaks of influenza in families, institutions, and other circumstances in which people are closely associated together, and from such observations they convince themselves that influenza is being transmitted from one person to another. All that is happening in such examples of tightly knit outbreaks, however, is that a patch of virus (as indicated schematically in Fig. 9.2) falls on the group in question, to produce an apparently connected outbreak of the disease.

Under controlled laboratory conditions, with volunteers inoculated by a threshold dose of influenza virus, the incubation time of the disease is about three days. Epidemics caused by person-to-person transmission would therefore be drawn out in time by the need for incubation to occur at each link of the transmission chain. A cloud of virus settling simultaneously on a community from the atmosphere would produce a sharper, more explosive outbreak of the disease, however. In principle therefore, the person-to-person transmission chain can be separated from the infall of a viral cloud, by observation of the times of onset of the disease in closely-knit communities. Such observations have been made and they favour markedly the infall of a virus cloud. Thus Stuart-Harris and Geoffrey Schild write:[1]

> '... studies in schools have shown that epidemics occur explosively in such a manner that it is difficult to explain the rapid build-up on a case-to-case transmission.'

From time to time we have been accused of 'violating Ockham's razor', evidently by people ignorant of the meaning of this phrase. The essential feature of Ockham's razor is not to choke off new

[1] Stuart-Harris, C.H. and Schild, G.C. *Influenza* (London: Edward Arnold, 1975), p.122.

ideas, but to stop old ideas from defending themselves against inconvenient facts by coining a series of *ad hoc* hypotheses, one hypothesis for each awkward fact. The hypothesis which has been coined to explain the observed explosive outbreaks of influenza in tightly-knit communities is that of a 'supershedder' of virus. A supershedder is an imaginary person who is supposed to carry a high concentration of virus without being particularly affected by it, as with a carrier of typhoid or cholera. In other words, a supershedder behaves within a community in exactly the way that the infall of a cloud of virus would behave.

In the early months of 1978, we obtained data on influenza outbreaks from more than 200 independent schools in England and Wales which proved that influenza is not spread by supershedders. Table 11.1 gives the absences from class of influenza victims,

Table 11.1. Class-absences of Boarders at Headington School, Oxford (classified according to House)

	House				
Date	*Davenport* (34 pupils)	*Hillstow* (63 pupils)	*Latimer* (46 pupils)	*Napier* (42 pupils)	*Celia Marsh* (44 pupils)
Jan 30	1	7	2	6	8
31	5	10	3	7	6
Feb 1	3	12	1	6	4
2	6	10	2	3	5
3	7	8	6	3	4
6	3	4	26	9	5
7	1	4	26	6	3
8	2	4	22	4	3
9	1	1	25	3	3
10	2	1	20	5	3
13	0	4	6	0	0
14	0	2	3	2	1
15	0	3	1	2	0
16	0	2	0	2	1
17	0	2	2	2	0

analysed with respect to house boarders at Headington School, Oxford. Plainly, something very drastic happened in Latimer House over the week-end of February 4 and 5. A small dense cloud of virus falling on Latimer during the evening or night of February 3, but not on the other houses, would explain these results. A supershedder in Latimer would not explain them, because during the daytime hours a supershedder would have mixed with pupils from the other houses.

If supershedders exist they should exude exceptional quantities of virus in school classes, where it will be essentially randomly distributed with respect to school houses. This will be true even if supershedders themselves succumb to the disease, since they will still exude the virus during incubation. Hence no school house can expect to escape infection by supershedders, regardless of which particular house the supershedders themselves are members. But school houses do escape infection, as for instance at Eton in 1978, where College House with 70 pupils had but a single victim in an epidemic with an average 35% attack rate for the whole school (441 victims out of 1248 pupils in the whole school). At another school where the two main houses each had about 55 pupils, one house had 2 victims and the other had 35 victims. Such cases, of which there are many, show the hypothesis of the supershedder to be plainly wrong.

One of the schools in our survey had conditions that were particularly well suited for testing the doctrine of case-to-case transmission. There were 48 victims among about 340 pupils, who all boarded under similar conditions. There were essentially 85 small dormitories, each with 4 pupils. The beds in each dormitory were close enough together to provide the readiest transfer of virus from pupil to pupil, if indeed infection goes at all in such a way. Thus case-to-case transmission would lead to the prediction of non-random clustering of the 48 victims in the 85 dormitories. There should be fewer dormitories with single victims in them than would be expected on a random basis, and more dormitories with clusters of two, three, and four victims in them. Calculation shows that a typical random situation would be 46 dormitories with no victim, 31 dormitories each with 1 victim, 7 dormitories each with 2 victims, 1 dormitory with 3 victims, and no dormitory with 4 victims. The actual results were as follows:

44 dormitories with no victim
35 dormitories each with 1 victim
5 dormitories each with 2 victims
1 dormitory with 3 victims
0 dormitories with 4 victims

The actual result was therefore very slightly less clustered than one would get in the most typical random distribution, but not by an amount with statistical significance. In effect, the situation was random, and there was no case-to-case transmission. It would be difficult to conceive of a more direct refutation of this incorrect belief.

A very similar conclusion has also been reported by R.E. Hope-Simpson from a survey of influenza attack patterns in domestic households. We discuss this work in our appendix. It would seem that there was little if any person-to-person transmission under living conditions of the greatest possible personal intimacy.

It would also be hard to conceive of a more decisive series of observational tests of the prediction that the influenza virus, or at least a component of it, arrives from space. The proofs are everywhere plain to be seen, in many different directions. The really surprising aspect of the situation is that the changes to the Arrhenius theory discussed in Chapter 8 should have led from Arrhenius' own pessimistic conclusion,

> 'There is little probability, though, of our ever being able to demonstrate the correctness (of the theory) ....'

to so many observational and experimental connections with the world around us. Sterility has been replaced by a wide perspective, which we shall continue to extend in the next chapter.

12
BACTERIA ON PLANETS

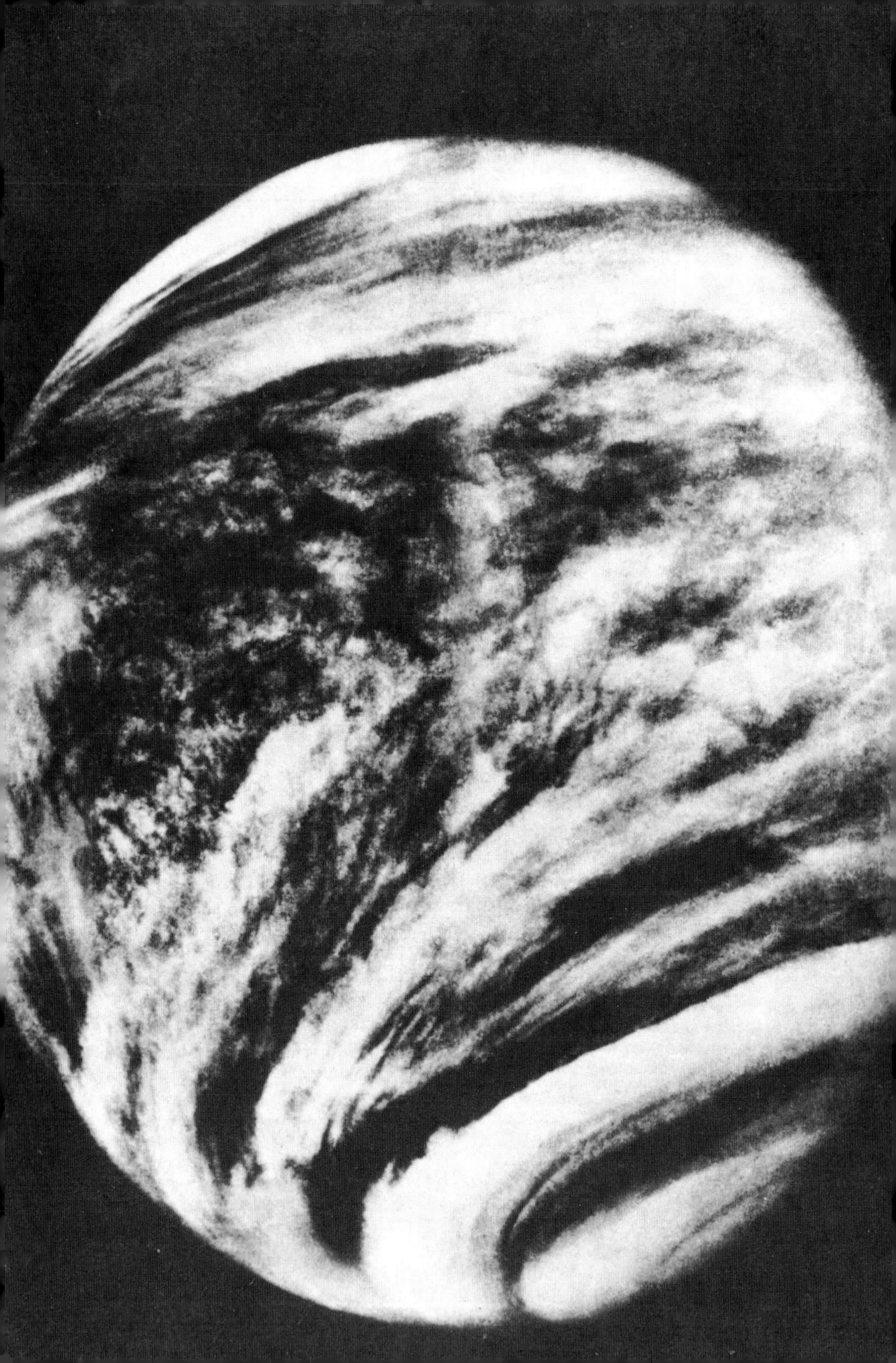

In this chapter we return to particular issues. Our picture has been of an immensely powerful universal biology that comes to be overlaid from outside on a planet such as our own. Wherever the broad range of the external system contains a life form that matches some local planetary habitat, the form in question succeeds in establishing itself. The external system contains a wide spectrum of choices which the local environment proceeds to select according to the particular niches that the local system provides. In our view the whole spectrum of life, ranging from the humblest single-celled life forms to the higher mammals (and beyond) must be overlaid on a planet from outside.

Bacteria exist everywhere on the Earth and in astonishing profusion. A handful of garden soil contains typically about a billion bacteria, and so do certain samples of deposits taken from the floors of the oceans. The seemingly hostile continent of Antarctica teems with bacteria — they are to be found in the soils and inside the rocks of dry valleys, and they exist and multiply towards the bottom of vast glaciers. In total, there are something of the order of 10^{27} bacteria on the Earth, of which only a minute proportion are pathogenic to man.

It is essential to all life-forms that they have access to energy. Photosynthesis, whereby sunlight provides the energy supply, is of course well-known, with chlorophyll playing a crucial role in the production of sugars from carbon dioxide and water. There is a widespread but erroneous popular view that photosynthesis is the basis of all life. Rather is it the case that the majority of bacterial forms derive their energy quite otherwise. Wherever an energy-yielding chemical reaction exists, and which proceeds only very slowly under inorganic conditions, some bacterium will usually be found to be living on it, speeding the reaction catalytically. There are bacteria that live by oxidizing sulphites to sulphates, which one might perhaps have expected on an oxidizing planet like the Earth. More unexpected are bacteria that use free hydrogen, an entirely evanescent substance on the Earth, to reduce sulphates to free sulphur. Nor would one expect to find the bacteria that use only free hydrogen and carbon dioxide to produce methane and water.

It is generally recognized that bacteria requiring free hydrogen cannot have evolved on the Earth over the last three billion years, and so they are supposed to be survivors from an earlier epoch in which free hydrogen is assumed to have been in plentiful supply. Yet the survival thread of these bacterial forms is so thin that it would surely be snapped in only a few centuries, or at most a few millennia. The clear implication is that the Earth receives a steady, continuing supply of sulphur and methane-producing bacteria, and that whenever unusual local conditions happen to set up an environment in which such bacteria can prosper they do so temporarily until the conditions change.

Modern industry with its dumps, tips, and tailings has created exceptional niches for bacteria which are always filled, even though those conditions never existed before on the Earth, except conceivably as a short-lived fluke. There is almost nothing that one can do in daily life which does not create a niche for some form of bacterium or protozoa. The farmer's bales of hay create a niche, and so does a bird's nest.

It is usual in biology to argue that bacteria and protozoa evolve genetically with great speed and so manage to adapt themselves to whatever possibilities the environment may offer. This view persists in spite of there being a considerable weight of evidence against it. Experiments in the laboratory over many generations have shown that bacteria do not evolve. Bacteria (and viruses) are incorrigibly stable. Experimenters can do three things to bacteria. Where several varieties are already present initially, one of them can be encouraged to multiply preferentially with respect to the others. Another thing that can be done to bacteria and viruses is to ruin them. The third alternative consists in dividing up what is already there and reassembling the bits in a different order, as for instance if one were to take selected bits from two different kinds of bacteria and reassemble them in an attempt to form a new bacterium. What experiments in the laboratory cannot do, however, is to make effective new bits. This is to be expected if the bits owe their origin to a galaxy-wide evolution which gave rise to a much higher order and more subtle system than is usually supposed.

True terrestrial evolution has consisted in the fitting of bits into aggregates that were optimal to the environment, as was demonstrated for plants and higher animals by classical biology. The bits have not been evolved to fit the environment, however. They are simply the inflexible cosmic bits that could manage to survive in the environment. The distinction shows clearly in the phenomenon of adapation. Birds, rodents, fish, and insects are clearly well-adapted to the environment in their different ways. Bacteria, on the other hand, are not closely adapted. Bacteria found in the ocean depths are not particularly adapted to those depths — they are simply bacteria that happen to function in a manner that is insensitive to high pressure. Many forms of bacteria are found surviving under temperatures that are far from optimal. Many found in warm soils multiply better at lower temperatures than are ever found in the tropics. Others multiply best at temperatures much higher than exist on the Earth at all, above 75 °C. There is indeed a very considerable measure of *dis-adaptation*, which shows itself in bacteria possessing properties that are irrelevant to survival on the Earth. Resistance to enormous doses of X-rays, ultraviolet light, and resistance to great cold were examples mentioned in former chapters. The cell walls of bacteria are stronger than reinforced concrete, and this is another example. Biologists are constantly being surprised by new unearthly properties, and the erection of almost any really novel environment reveals some such property. The canning of food provided a new environment. There was a practical need to find a cheap and convenient means of sterilizing food after its sealing inside a metal can. Flooding by X-rays was tried, and it was then that *Micrococcus radiophilus* was discovered. Heating was also tried, and some bacteria were found to be capable of surviving temperatures above 100 °C. Water in the open expands slightly as it is heated. Water in a closed can of fixed shape cannot expand, however. Instead a high pressure is created, which can prevent the formation of bubbles of steam even above 100 °C. Likely enough it was the prevention of steam bubbles which permitted bacteria to survive in these experiments. But how could such a property ever have evolved on the Earth?

Manifestly, bacteria are a life-form that has simply been plastered over the Earth, to survive wherever it is possible to do so. The properties of bacteria have a galaxy-wide quality, much wider than is needed for survival on the Earth alone. Indeed, because of the great breadth of their survival characteristics, the potentiality of bacteria to establish themselves on planets must be much greater than one would permit oneself to suppose within the narrow confines of an Earth-bound theory of the origin of life. Just because the other planets of our solar system are very different from the Earth is no reason at all for thinking that life in our system must be unique to the Earth. The spread of the survival characteristics of bacteria are so great that they may well include the ability to fit environmental possibilities on some other planet or planets. In this chapter we shall find suggestive evidence that this is actually so for Venus, Jupiter and Saturn, and that it is possibly so for Mars, Uranus, and Neptune.

We take as an indispensable requirement for the survival of bacteria that water can condense on them, and can pass to their interiors and be held there in a liquid form. This condition rules out Mercury and the Moon. The lifeless states of Mercury and the Moon show immediately in their drab lack of colour. As a valid general correlation, life and colour may be associated together, although perhaps not in what mathematicians call a one-to-one correspondence. A stricter indication of life comes when particle sizes close to 1 μm turn out to be dominant, in which connection we recall the size distribution of spore-forming bacteria given in Fig. 7.2.

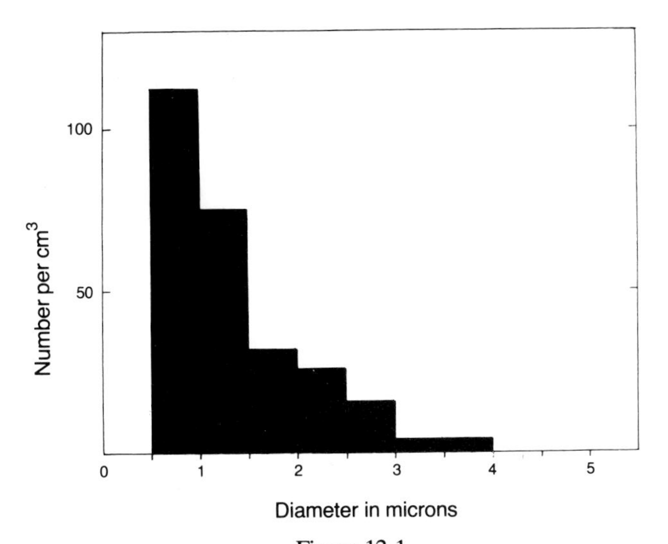

Figure 12.1
Distribution of particle sizes in the upper clouds of Venus (approximately 55–65 km in altitude) measured by Pioneer Venus.

The upper clouds of Venus produce a rainbow, indicating that the cloud particles are spherical and that they have sizes in the region of 1 μm[1]. The distribution of particle sizes actually measured in the upper clouds by Pioneer Venus is shown in Fig. 12.1.[2] Although a considerable amount of further argument is needed before one can be satisfied about details, the correspondence of the observed particle sizes with those of bacteria tells the main story immediately.

There are two ways in which bacteria can be spherical in shape, one through forming spores and the other through restriction to the class of micrococci. Of these, spores appear the more likely possibility, for the reason that the clouds of Venus are in convective motion, extending from an upper level at an altitude of 65–70 km

[1] Coffeen, D.L., and Hansen, J.E., *Planets, Stars and Nebulae Studied with Photopolarimetry,* ed. T. Gehrels (University of Arizona Press, 1974).

[2] Knollenberg, R.G., and Hunten, D.M., *Science* 203 (1979), 792.

down to an altitude of 45 km. The temperature at 45 km is about 75 °C and at the top is about −25 °C. While survival over this range is easily possible for bacteria, the repeated variations of temperature caused by a circulating cloud system would be better resisted by bacteria capable of forming spores which are still more hardy than the bacteria giving rise to them. Thus bacteria could be rod-shaped in the lower warmer regions, but giving place to spherical spores in the cold upper regions of the clouds. In this way there is no requirement for the bacteria to be exclusively micrococci.

It was pointed out in Chapter 7 that a quantity called the 'refractive index' of a particle is relevant to its light scattering properties. The refractive index of water is about 1.33 and that of biological material about 1.5. Bacterial spores contain about 25% water by weight, which causes them to behave like a uniform particle with refractive index 1.44, which is exactly what is shown by observations on the degree of polarization of the upper clouds of Venus.

The refractive index of sulphuric acid is 1.48, and sulphuric acid droplets with 25% water would also behave like uniform particles with refractive index 1.44. But this explanation of the upper clouds of Venus fails on a number of important points. There is no reason for such droplets to have sizes of 0.6–1.2 μm (Fig. 12.1). Sulphuric acid droplets could, and very likely would, have a much broader size distribution than Fig. 12.1. The 25% water concentration is an arbitrary choice, whereas for bacterial spores there is no choice, about 25% is what the water concentration must be. The chemical sampler carried by Pioneer Venus (1978) measured vapour pressures of oxygen and sulphur dioxide in the cloud regions that were thousands of times lower than one can easily have in the laboratory, and yet, even at the much higher laboratory concentrations, oxygen and sulphur dioxide do not go easily to sulphur trioxide, and thence by the addition of water to sulphuric acid. Commercial sulphuric acid is produced in two ways, the old way by a complex of reactions involving nitric acid, and the modern way by catalytic processes involving either platinum metal or vanadium oxide. Inorganic catalysts like platinum tend to become 'poisoned' under natural conditions, however — they take part in other reactions which change them. The highly effective catalysts found in nature are bacteria, which not only maintain themselves (which is all that inorganic catalysts can do) but actually increase in number through the chemical reactions which they promote. Hence to produce some sulphuric acid, which is apparently required to explain certain details of the infrared radiation emitted by the clouds of Venus, it is to bacteria that we should look. Sulphur bacteria are yellow in colour, and it is to their presence (among other non-yellow bacteria) that we attribute the pale yellow colour of the light reflected by Venus. Sulphuric acid is colourless, on the other hand, and droplets of it would not produce any such colouring.

Water is not a particularly abundant constituent of the atmosphere of Venus. The second atmospheric sample taken by Pioneer Venus at an altitude of about 46 km near the base of the lower clouds gave about 0.5% water. With this concentration, and choosing −10 °C as the lowest temperature at which the bacteria replicate, the relative humidity can be calculated to be about 85%, which is adequate for bacteria with an appreciable internal content of dissolved salts. It is important here that the problem of maintaining liquid water inside a bacterium is not simply evaporation from a free water surface but of evaporation from the outer surface of the bacterial membrane, which is markedly water-attractive. At 85% relative humidity only a quite small extra bit of holding power against the evaporation of water molecules is sufficient to stop a bacterium from drying out in the manner of Fig. 7.3. However, under exceedingly dry conditions, near zero per cent humidity, bacteria must largely dry out. As bacteria circulated through the middle and lower cloud regions of Venus, they would be subject to relative humidity values that ranged from near zero in the hotter, lower regions to about 85% in the cooler, upper regions. This would provide a natural water pump with alternating phases of filling with water and of subsequent evaporation, and with each water-filling episode giving a fresh supply of nutrients to the interior of the cell.

Since Venus is exceedingly hot at ground-level (about 450 °C) it is a matter of some surprise to find evidence of life existing there. The circumstance which makes life possible is the dual circulatory pattern of the Venusian atmosphere. From the temperatures and pressures measured by Pioneer Venus one can infer the presence of a lower convective zone from ground level to a height of about 30 km. There is then still atmosphere up to about 45 km, where the second convection zone begins. It is the non-moving in-between region from 30 km to 45 km which protects life in the higher zone from being quickly swirled down to the impossibly hot conditions near the ground.

Although bacteria are small and are able to ride easily with the atmospheric motions, there must nevertheless be occasional situations where a bacterium carried down to the base of the higher convective zone fails to find an up-current on which it can ascend again. Inexorable gravity will then cause the bacterium to fall slowly down through the in-between zone until it reaches the lower convective zone, where it will quickly be snatched downward and destroyed by the heat. It is therefore to be expected that the in-between zone will contain a thin haze of slowly falling doomed bacteria, and the existence of a thin haze between altitudes of 30 km and 45 km was indeed found by Pioneer Venus.

From the data obtained by Pioneer Venus one can infer that the fall-out through the in-between zone would begin to denude a population of high-level bacteria in about 30 000 years, if the bacteria did not renew themselves. For renewal a supply of nutrients is required. Water, nitrogen, and carbon dioxide are amply available, but in addition to these main ingredients other elements — sodium, magnesium, phosphorus, sulphur, chlorine, potassium, calcium, manganese, iron, cobalt, copper, zinc, and molybdenum — are needed in smaller proportions. It can be calculated that the fall-out through the in-between zone would carry about 10 000 tonnes of these essential ingredients each year from the higher convection zone to the lower zone. How, one can ask, may such a loss from the higher zone be compensated? The answer appears to be by the infall of meteoric material from space, rather than by gas diffusion upward from the lower atmosphere. The meteoric supply to the Earth is known to be about 10 000 tonnes per year, and the supply to Venus — a very similar planet — should be about the same. The problem for gas diffusion upward lies in the lack of volatility of the compounds of many of the needed elements. There is no such problem for meteors in the size range from about 0.1 mm to 1 cm, which are gasified as shooting stars on plunging at great speed into the high atmospheres of the Earth and of Venus.

The size distribution of particles measured for the upper clouds of Venus is essentially maintained in the middle and lower clouds, and in the haze zone of the in-between region. In the middle and lower clouds, however, there is also a considerably less numerous population of much larger particles. These we attribute to the tendency of bacteria to aggregate into colonies, a property that may well be helpful in preventing too much evaporation of water in the dry, hot conditions of the lower clouds.

The correspondence between the rate of supply of meteoric material and the rate of loss of essential nutrients down through the in-between region, about 10 000 tonnes per year in each case, suggests that the quantity of bacteria in the clouds is nutrient-limited. The clouds have built up until the drop-out at the bottom equals the supply from outside. If the supply from outside were to cease then the bacterial population would decline steadily, until after a few tens of millennia it would be possible to see down to the ground-level of Venus. Such a situation may well have occurred in past times, only then there were no humans to observe the lower excessively heated surface zone of the planet.

Humans have for long looked at the red colour of Mars and seen evidence there for the existence of life. The argument was that the red colour implies a highly oxidized condition, a conclusion that a recent Mariner landing on Mars has shown to be correct. The supply of oxygen needed to produce this condition might have come from photosynthetic organisms, and this further aspect of the old argument may also be correct. But if so the organisms must have existed in the remote past when liquid water was present at the surface of the planet. The Martian surface is cut by many sinuous channels, of which that in Fig. 12.2 is an example. It is generally agreed that these channels were made by a liquid much less viscous than molten lava, and water is a likely possibility. But there is no liquid water nowadays at the surface of Mars, and consequently (if we take the need for liquid water to be essential) there can be no active surface life. It is surprising therefore that so great an effort was made in the recent Mariner landing to look for evidence of life. It is true that bacteria in a dormant condition might survive for long periods in the Martian surface dust, but such bacteria might well need unusual conditions for growth, which could not be anticipated in a terrestrially designed experiment.

If one argues analogously to the Antarctic, the best chance for life to be active on Mars would be deep inside glaciers, where the temperature might rise sufficiently for water to become liquid. There would still be problems of nutrient supply, but if the glaciers themselves turn over, top to bottom from time to time, such problems would be capable of solution. The bacteria would need to live on some energy-producing chemical reaction, and, if the reaction had a gaseous product (such as carbon dioxide or methane), the possibility would exist for the building up of subsurface pockets of gas, which might explode sporadically to the surface unleashing quantities of bacteria, spores, and inorganic dust into the Martian atmosphere. In this connection we recall the vast atmospheric dust storm which greeted the arrival of a Mariner vehicle in 1971. This storm has been attributed to high winds generated by the normal Martian meteorology, but if so one might

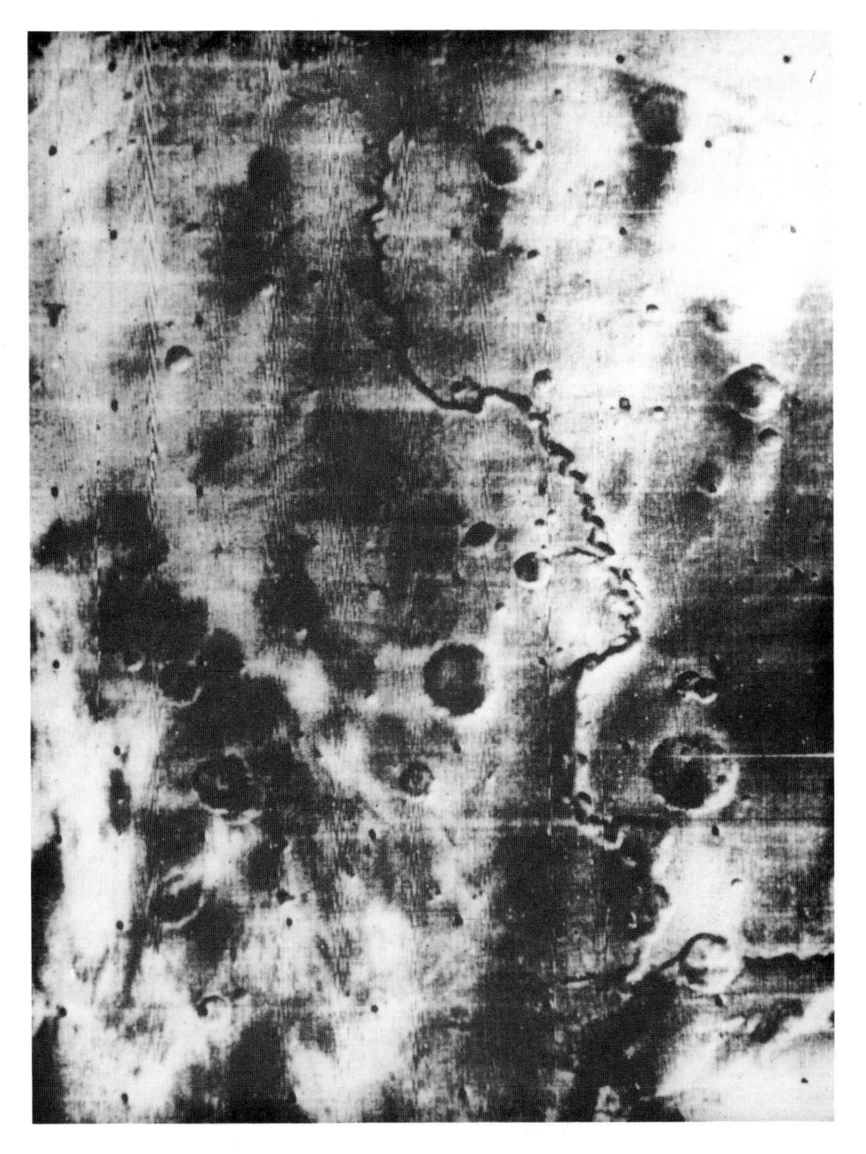

Figure 12.2
A sinuous channel on Mars, probably produced by the flow of liquid water. (Courtesy of NASA.)

wonder why such winds are not a seasonal phenomenon. The outward explosion of microbiologically-generated gas could prove to be a better explanation.

From an analysis of reflected sunlight it has been estimated that the clouds of Jupiter consist of particles with diameters close to 0.5 μm and with refractive index 1.38. This diameter value is somewhat less than the main peak of the distribution of Fig. 7.2 at about 0.7 μm. The calculations for Jupiter were made, however, on the assumption of spherical particles. Since Jupiter does not show a rainbow effect like Venus, it is to be doubted that this assumption of sphericity is correct, and for rod-shaped particles the calculation would be somewhat changed. The difference between the calculated 0.5 μm and the 0.7 μm of Fig. 7.2 does not seem, therefore, to be a barrier to the particles of the Jovian clouds being interpreted as bacteria. Unlike bacterial spores, which contain about 25% water, rod-shaped bacteria, when active, contain about 70% water by weight. Estimating the equivalent refractive index of a uniform particle along the same lines as before, we obtain 1.37, which is close to the value 1.38 estimated from the reflecting properties of the clouds of Jupiter.

Water is not a problem on Jupiter. Much of Jupiter's oxygen is combined with hydrogen into water, and in the lower atmosphere below the visible upper clouds liquid water drops probably exist in profusion.

We have emphasized already that bacteria are able to operate in almost any sense in which energy can be released by chemical reactions. Their many varieties simply take what is available . But if Jupiter is considered as a closed chemical system there can be very little that is available. In the absence of photosynthesis, which seems unlikely to take place low down in the clouds (except possibly under an exceptional condition to be considered later) degradable materials are not produced continuously within the system, and any degradable materials there may have been in the beginning would long ago have been used up. So Jupiter can operate as a biological system only if degradable materials are supplied from without. Injected into the atmosphere of Jupiter, most meteoric material would, after vaporization, be degradable in this sense. Metallic iron is oxidizable in the presence of water, and there are bacteria which operate precisely on this transformation. As we have already remarked, sulphates are reducible to sulphur, and carbon dioxide is reducible to methane.

The flux of meteoric material in the neighbourhood of Jupiter has been measured to be about 100 times greater than it is near the Earth, due to a focusing effect produced by Jupiter's gravitational field. Moreover, the surface area of Jupiter is about 100 times greater than that of Earth. It follows therefore that the incidence of meteoric material onto Jupiter must be some 10 000 times greater than the incidence onto the Earth. Since the latter may be as high as ten thousand tonnes per year, the incidence onto Jupiter may well be 100 million tonnes per year. The energy produced by the chemical degradation of such a quantity of material is comparable with, but rather less than, the energy to be obtained from the burning of a similar quantity of wood. While very considerable, this amount of energy is smaller than the energy involved in photosynthetic processes on the Earth (the amount of wood grown annually is much more than 100 million tons). It must be remembered, however, that terrestrial biology operates in an oxidizing environment, which forces a high metabolic rate, otherwise the biological system would soon burn up. No such tendency to burn up exists on Jupiter. The great quantity of free hydrogen in the Jovian atmosphere removes all oxygen, and bacteria could survive there in a dormant condition for a very long time. Destruction rates would be low and populations could gradually be built up over extended periods. The ochre, red, and brown belts and spots of Jupiter can be identified with sulphur and iron bacteria.

It is one of the remarkable features of meteoric material that all size ranges contribute appreciable quantities of mass. This is true even for large bodies several kilometres in radius. It has been estimated that two or three such bodies hit the Earth in a million years, and for Jupiter there would correspondingly be two or three large bodies in a century, each contributing about 100 million tons of nutrient material in one sudden burst. A kilometre-size object hitting Jupiter's atmosphere at high speed would be disintegrated into heated gas that would be sprayed out over a considerable area, but still over an area that was small compared to the vast total area of Jupiter. It would form a spot, a spot in which the supply of nutrients for biological activity was exceptionally high, so that a large bacterial population would be built up in the resulting area. It is possible in this way to understand the origin of the spots of Jupiter, of which the Great Red Spot, shown on page 6, is the best-known example.

The possibility exists for a feedback interaction to be set up between the properties of a localized bacterial population (for example, the infrared absorption and emission properties of the bacteria) and the general meteorology of Jupiter's atmosphere. If bacterial populations on Jupiter have become adapted to the meteorology there, it is possible that evolution has produced a situation in which populations are able to prevent supplies of nutrient materials from being swept away from them by atmospheric motions. In a measure they may have become able to control the meteorology, and thus to hold together spot concentrations of nutrient materials. This may well be the explanation of the persistence of spots, and in particular the remarkable persistence of the Great Red Spot.

Let us now turn to meteoric materials in the form of tiny sub-micron particles. These too must make an appreciable contribution to the supply of nutrient material. Such particles are always electrically charged by sunlight, and being of small mass they are subject to deflection by the strong magnetic field of Jupiter, which causes them to rain down on the polar regions of the planet rather than on the equatorial zone. Bacterial activity resulting from this specifically polar accretion of fine meteoric particles may well be responsible for the remarkable dappled appearance of Jupiter at its poles.

At first sight one might think that photosynthesis would be impossible on Jupiter. At the upper cloud levels where sunlight is available the temperature is probably too low to permit bacterial activity, and at lower levels where the temperature rises to appropriate higher values there can be little penetration of sunlight. In the neighbourhood of spots the clouds are by no means smoothly layered, however. If sunlight can ever penetrate far enough to reach places where water is liquid, photo effects become possible. We remarked above that without sunlight, and in the presence of free hydrogen, bacteria can reduce sulphates in the presence of water. With sunlight, however, both reduction and oxidation become possible. A wider range of bacterial colours then arises, with the addition of purple, blue, and green to the yellows, reds, and ochres of the reducing bacteria. Such a situation may not be inconsistent, with recent observations from Voyager I. Occasional blue colours are seen on Jupiter, suggesting that the special conjunction of sunlight and liquid water may sometimes arise there.

Astronomical observations, particularly the most recent Voyager I data, show a generally similar situation with respect to colour on the main body of Saturn, including the presence of spots although with more yellow and green than Jupiter. The well-developed polar coloration and the absence of exceptionally large spots suggests that meteoric additions to Saturn may well be much more in the form of a rain of tiny particles than as large spot-forming bodies.

The rings of Saturn seem also to contain large quantities of bacteria-sized particles. Spectacular pictures of the ring system relayed from cameras on Voyager I in November 1980 revealed many surprising features. The so-called Cassini division between the two outer bright rings, the A- and B- rings, was seen to be populated by many rings of fine particulate material. The Saturnian ring system, which has been likened to the grooves on a gramophone record, is shown in Fig. 12.3.

Two aspects of these recent observations are particularly noteworthy. Several spoke-like structures have been found to appear in the B-ring — Saturn's brightest ring seen from the Earth (Fig. 12.4). These spokes appeared as dark tongues in images that Voyager I relayed whilst being above the rings on their sunward side; but as the spacecraft went below the rings the same structures appeared as bright tongues. From this single observation we can infer that the particles in the tongues have the dimensions of bacteria. Such particles scatter sunlight mainly in a forward direction, so that they appear bright when the Sun is behind them, but dark when the Sun is in front. Another baffling feature of the new data concerns the braided and twisted appearance of the outer F-ring, which is also made up of bacterial-size particles (Fig. 12.5). Since two small icy satellites have been discovered in orbits that straddle the F-ring, a possible resolution of this peculiar ring structure is that its particles are currently being spewed out from the two icy satellites designated S13 and S14. These satellites, which have sizes appropriate to large comets, may have become trapped in the gravitational field of Saturn and been melted in their interiors due to tidal effects, or due to the effect of collisions with smaller objects. Such satellites would act like biological pressure cookers, releasing jets of microroganisms which had been produced in their melted interiors. We envisage a situation where the F-ring is woven *in situ* from jets of material thus released from S13 and S14. Magnetic effects, which involve the coupling of electrically charged bacteria with the planetary magnetic field, could play a role in maintaining both the twisted structure of the F-ring and the structure of the radial tongues seen in the B-ring.

No observations are yet available to determine the sizes of the particles which constitute the clouds of Uranus and Neptune. The atmospheres of these two planets are known, however, to contain

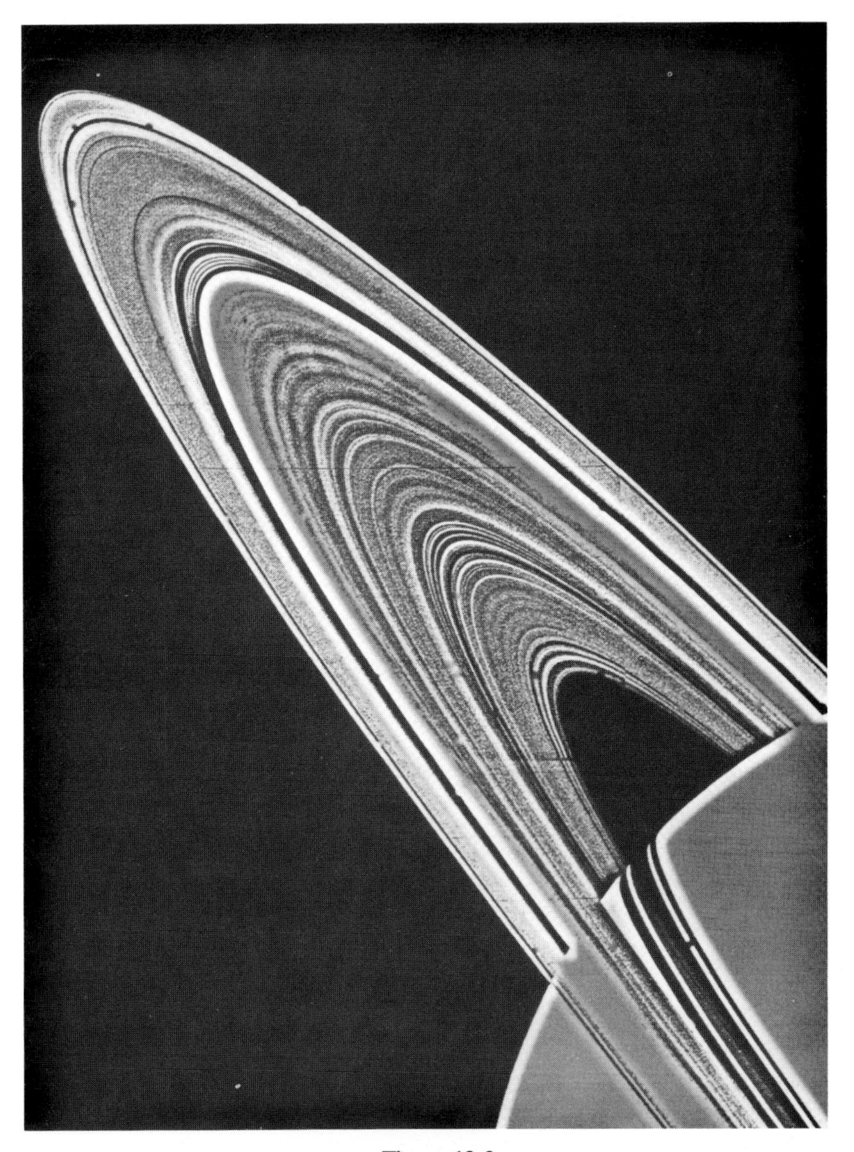

Figure 12.3
Saturnian ring system photographed by Voyager I. (Courtesy of NASA.)

Figure 12.4
Rings of Saturn showing radial tongues in the B-ring. The tongues are comprised of bacteria-sized particles. (Courtesy of NASA.)

Figure 12.5
The outer F-ring comprised of bacteria-sized particles has a braided structure. (Courtesy of NASA.)

vast quantities of methane, as do Saturn to a lesser extent, and Jupiter to a still smaller extent. The existence of methane may also be indicative of bacterial activity. Astronomers have been content to accept the methane observed in the atmospheres of the four large outer planets of the solar system as the outcome of a 'thermodynamic trend' of carbon compounds to go to methane at low temperature and in the presence of an ample supply of free hydrogen. But if one takes a vessel containing a mixture of free hydrogen and carbon monoxide or carbon dioxide the separate gases persist unchanged for an eternity. The 'trend' is so slow as to be essentially zero. It is in such circumstances that catalysts are used in the laboratory and in industry. Of all catalysts, bacteria are the most efficient. Indeed the methane-producing bacteria have evolved precisely to speed the conversion of carbon dioxide and hydrogen into methane and water. If the conversion happened at all readily under inorganic conditions there would be no niche for these bacteria. The natural explanation for the quantities of methane present in the atmospheres of Jupiter, Saturn, Uranus, and Neptune is that methanogenic bacteria have been active there in the reduction of carbon monoxide and carbon dioxide. The scope for such bacteria on the outer planets is clearly enormously greater than it is on the Earth, where this class of bacteria has been able to establish only a small toe-hold. Such bacteria could be responsible for the white zones of Jupiter, and for the generally uncoloured appearances of Uranus and Neptune.

13
BIOLOGY OR ASTRONOMY?

For an externally impressed biological system to gain a toe-hold in a new environment, as for instance bacteria which are incident from space upon some new planet, it is essential that the external system be already equipped for survival in the new environment. Once a toe-hold becomes established, however, a two-way adaptation can develop. The basic components (genes) of the biological system could become selected to match the environment, or the cosmically-imposed components could be built (without being much changed individually) into structures better suited to the environment. In both of these cases it is the biological system that makes changes in response to the restrictions imposed by the physical aspects of its surroundings. This is one direction in which adaptation can take place. The other direction is through the biological system altering the environment itself. It is this second way with which we shall be concerned in this final chapter.

Terrestrial biology has changed the composition of our atmosphere, it affects the break-up of rocks into soil, it affects the rates of run-off of surface waters and of evaporation from the land, and it even influences in some degree the Earth's climatic zones. We similarly expect biology elsewhere to influence local conditions. The meteorology of the atmosphere of Jupiter was mentioned in the preceding chapter as a possible example, and perhaps the meteorology of Venus may also be affected. These are minor effects, however, compared to what biology may be capable of achieving on an astronomical scale.

It is interesting that issues that would influence the well-being of life on a cosmic scale turn out to involve problems which have been under active consideration in astronomy for more than a generation, and for which no satisfactory inorganic solutions have been found. These issues involve the star formation process and its relation to the interstellar grains. The unresolved issues are as follows:

What decides the rate at which stars form from the interstellar gas?

When they are formed what decides the mass distribution of stars?

What decides the rotations of the stars, and whether they are formed with planetary systems?

What decides how star formation is correlated throughout a whole galaxy, leading to the production of both grains and stars on a far-flung basis, often with the appearance of a new set of spiral arms for a galaxy?

The answers to these questions are almost certainly connected with the existence of a magnetic field everywhere throughout our galaxy. But the nature and origin of the galactic magnetic field is a further unresolved problem, and so the additional question must be added:

How did the magnetic field of our galaxy come into being?

This further question has proved so baffling that many astronomers have given up hope of answering it, by claiming the magnetic field to be truly primordial, it being imposed on the Universe at the moment of its origin. The magnetic field is what it is, because it was what it was, right back to the first page of Genesis. If this view is correct the situation is crude, uncontrolled, and unsatisfactory.

To understand the difficulty of the problem, suppose we were to try to generate the magnetic field of our galaxy with the aid of an enormous electric battery, with one terminal connected to the centre and the other terminal to the outside of the galaxy. With the battery switched on, an electric current would begin to flow through the interstellar gas. Owing to a phenomenon known as 'inductance', the electric current would at first be exceedingly small, but as time went on the current would become stronger, and as it did so the magnetic field associated with the current would increase in its intensity. Suppose we allow the current to grow for the whole age of our galaxy, about 10 000 million years. What voltage do we need for the battery so that after 10 000 million years the resulting magnetic field will be as strong as the galactic magnetic field is actually observed to be? The answer is about 10 000 000 million (10^{13}) volts. What process, we may ask, could have produced a battery of such enormous voltage that could operate for a time as long as 10 000 million years? A conceivable answer is a stream of electrically charged interstellar grains projected at high speed, 100 km s^{-1} or more, into an electrically neutral gas, a process in some respects similar to that which drives a terrestrial thunderstorm. The required projection speeds of 100 km s^{-1} are attainable by radiation pressure, as we discussed in earlier chapters. The remaining problems are first to attain systematic directivity for a stream of grains, and second to maintain electrical neutrality in the gas through which the stream passes (when neutrality fails in a terrestrial thunderstorm there is a lightning flash, and the electric battery is instantly dissipated). With inorganic grains it is difficult, if not indeed impossible, to resolve these issues. Bacteria, on the other hand, have far more complex properties than inorganic grains, and may be able to exert a control both on stream directions and on the neutrality of the interstellar gas. The issue is not proved, but it is conceivable, and if it were to happen, bacteria would be well-placed to control the whole process of star formation.

The nutrient supply for a population of interstellar bacteria comes from mass flows out of the large galactic population of old stars (100 000 million of them), which may well have had an inorganic origin. 'Giants' arising in the evolution of such stars experience a phenomenon in which material containing nitrogen, carbon monoxide, water, hydrogen, helium, some refractory solid particles and supplies of trace elements flows continuously outward into space. In total from all giant stars, a mass about equal to the Sun is expelled each year to join the interstellar gas. This is the nutrient supply.

The problem for interstellar bacteria is that the nutrient supply cannot be converted immediately into an increase of the bacterial population, because of the need for liquid water, which cannot exist at the low pressures of interstellar space. Water in interstellar space exists either as vapour or as solid ice, depending on its temperature. Only through star formation, leading to associated planets and smaller bodies, can there be access to liquid water. Conditions suited to the presence of liquid water can exist over long periods of time on planets like the Earth. Liquid water need not exist, however, for long periods of time, since bacteria can multiply so extremely rapidly given suitable conditions. Shorter periods could exist on bodies much smaller than planets, and in the early high-luminosity phase of newly formed stars the bodies could lie far out from the stars, at much greater distances than the Earth is from the Sun. In the case of our own solar system, liquid water could quite well have existed in the early days far out towards the periphery, and it could have existed at the surface of bodies of lunar size or inside still smaller bodies. The nutrients present in the outer regions of the solar system must have exceeded by many millions the amount at the Earth's surface. Hence the short-lived conditions associated with star formation must be of far greater importance to the population of interstellar bacteria than the long-lived, more or less permanent environment provided by planets like the Earth.

It has long been clear that the detailed properties of our own solar system are not at all what would be expected for a blob of interstellar gas condensing in a more or less random way. Only by a very strict control of the rotation of various parts of the system could such an arrangement as ours have come into being. The key to maintaining control over rotation would seem to lie once again in a magnetic field, as indeed does the whole phenomenon of star formation. The surest way for interstellar bacteria to prosper in their numbers would be through maintaining a firm grip on all aspects of the interstellar magnetic field. By so doing they would control not only the rate of star formation but also the kinds of star systems that were produced.

The multiplying capacity of bacteria is enormous, as we pointed out in earlier chapters. To go from an individual bacterium to the number of all the interstellar grains requires about 170 doublings. When conditions are optimal a bacterial population can double in a few hours, so 170 doublings take less than a month. Of course such a prodigious explosion in number would never literally be achieved because of practical limitations occurring in the availability of nutrients. Nevertheless, there is evidence that whole galaxies are overwhelmed from time to time by comparatively rapid and very large scale episodes of grain formation, as for instance the galaxy M82 shown in Fig. 5.5. This example is far from unique. There are many cases of galaxies embedded in a vast cloud of particles. On a lesser scale, there is a similar distinction between two appendages to our own galaxy, the Large and Small Magellanic Clouds.

For a generation or more astronomers have been accustomed to thinking of star-forming episodes accompanied by the production of vast clouds of interstellar grains. The episodes are sometimes local but they are often galaxy-wide. They are thought to be triggered by some large-scale event, the after effects of which linger on for some considerable time, several hundred million years. The condensation of the exceptionally bright stars which delineate the spiral structures of galaxies has often been associated with these episodes. From our argument it seems then that even the origin of the spiral structures of galaxies may well be biological in its nature.

The potential of bacteria to increase vastly in their number is enormous. It should occasion no surprise, therefore, that bacteria are widespread throughout astronomy. Rather would it be astonishing if biological evolution had been achieved on the Earth alone, without the explosive consequences of such a miracle ever being permitted to emerge into the Universe at large. How could the Universe ever be protected from such a devastating development? This indeed would be a double miracle, first of origin, and second of terrestrial containment.

Some biologists have probably found themselves in opposition to our arguments for the proprietary reason that it seemed as if an attempt were being made to swallow up biology into astronomy. Their ranks may now be joined by those astronomers who see from these last developments that a more realistic threat is to swallow up astronomy into biology.

Appendix
A Technical Note on Influenza

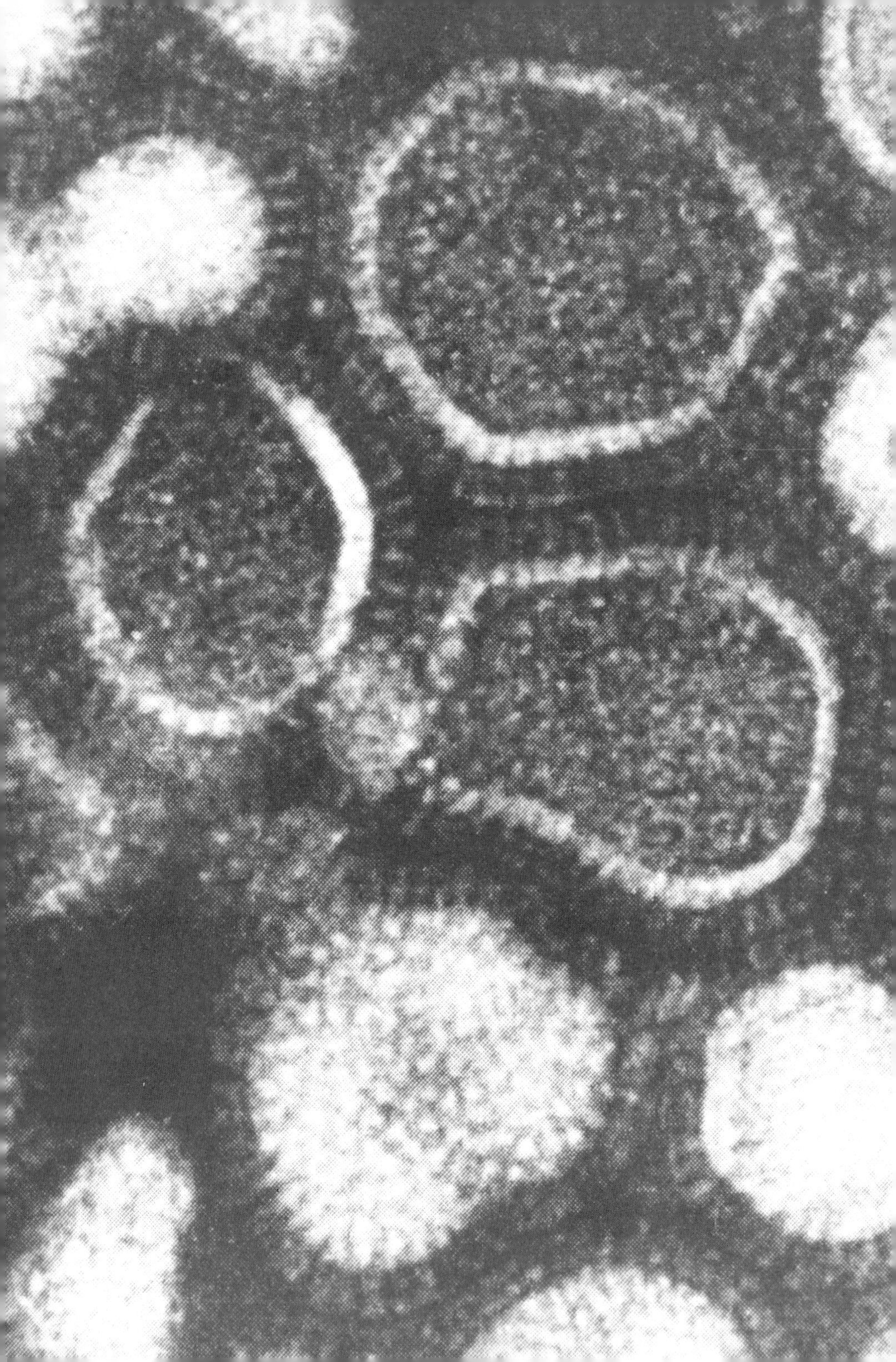

Introduction

Suppose a well-defined laboratory strain of influenza virus, V_1 say, is used to infect an experimental animal. Let the output of virus, V_2 say, be used to infect a second animal, and let the output of virus from the second animal, V_3 say, be used to infect a third animal .... and so on, thereby generating a sequence of case-to-case transmissions under controlled laboratory conditions. It is found in such a series that:

$$V_1 \simeq V_2 \simeq V_3 \simeq \ldots \simeq V_n.$$

The final output virus V_n is essentially similar to the initial input virus V_1. Yet, if such a series is generated with the first case coming from a virulent natural infection, it is found that V_n lacks the virulence of the initial infection.

It has been usual to suppose that for a series initiated by a natural infection, the virus changes in some way, perhaps due to copying errors along the chain, but this is contrary to other evidence which shows that genetic material is copied in cells with very high efficiency. It is also contrary to what is believed to obtain in pandemic situations where V_n is supposed to maintain its virulence even though the sequence of case-to-case transmissions is very long.

One way to escape from this paradox is to drop the implicit assumption that natural influenza is caused by a fully-fledged virus. To examine this possibility, we choose a more general model,

$$A_1 + A_2 + \ldots \rightarrow \text{Animal} \rightarrow \text{Virus},$$

where A_1, A_2, ... are activation factors that need not be fully-fledged virus. Some of the factors are to be regarded as essential if the disease occurs, but others may be optional.

As a particular case of this general model we discuss the possibility:

$$A_1 + A_2 + A_3 \rightarrow \text{Animal} \rightarrow \text{Virus},$$

where A_1 is a triggering agency (a viroid), with the genes of the output virus being contained in either A_2 or A_3. The reason why A_2 and A_3 are presented as alternative sources of the viral genes will appear at a later stage. The first essential is to convince oneself of the need for the trigger A_1 and that A_1 must be incident on the Earth from space.

Reasons why there must be at least one world-wide activation factor

Before Pasteur's work showed microbes to be the agents whereby diseases and other forms of biochemical activity are transmitted from place to place, the strange epidemiology of influenza had suggested that conditions in the Earth's atmosphere controlled the incidence of the disease. In his book *The Modern Practice of Physic*, published in 1813, Robert Thomas commented:

> 'By some physicians influenza was supposed to be contagious; by others not so; indeed, its wide and rapid spread made many suspect some more generally prevailing cause in the atmosphere.'

Pasteur's work demonstrated in the opinion of most doctors and biologists that influenza had to spread by contagion. Although the facts of epidemiology remained just as cogently against this opinion as they had been in the time of Robert Thomas, consciences were salved by the suggestion that, if one looked for long enough in the opposite direction, the facts would somehow go away. They have not done so of course.

On the contrary, a large body of data has accumulated over the years that decisively goes against the doctrine of contagion. Some of this data has already been discussed in Chapter 11. We shall now consider further evidence for the occurrence of at least an activation factor for influenza on a global scale.

Figure A.1 gives data obtained by R.E. Hope-Simpson (*J. Hyg. Camb.*, 83 (1979), 11). The histogram on the left shows negligible person-to-person transmission within 53 households in 1968–69

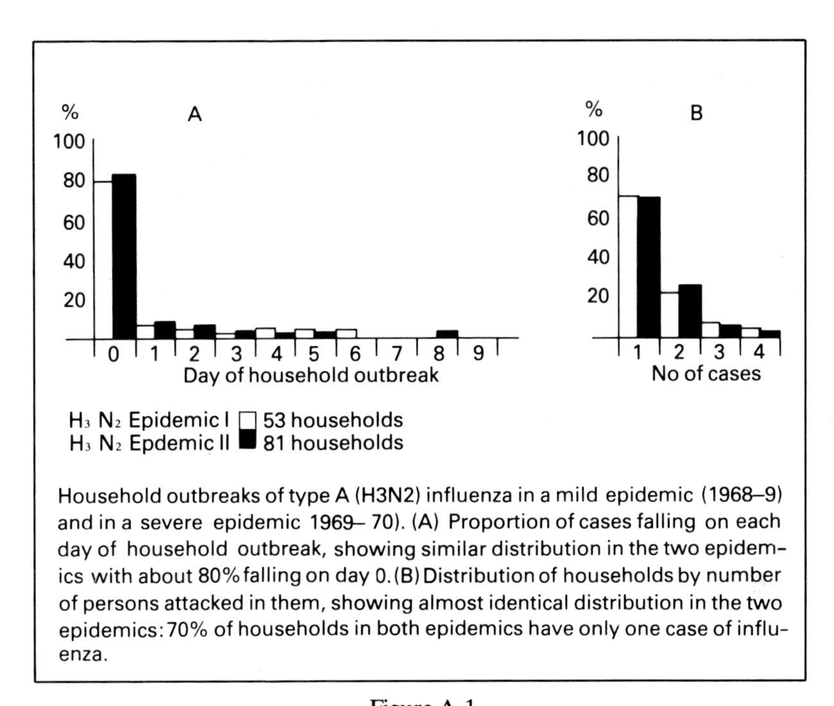

Household outbreaks of type A (H3N2) influenza in a mild epidemic (1968–9) and in a severe epidemic 1969–70). (A) Proportion of cases falling on each day of household outbreak, showing similar distribution in the two epidemics with about 80% falling on day 0. (B) Distribution of households by number of persons attacked in them, showing almost identical distribution in the two epidemics: 70% of households in both epidemics have only one case of influenza.

Figure A.1
Data due to R.E. Hope-Simpson on the incidence of influenza within households during two epidemics showing the lack of significant person-to-person transmission.

(open rectangles) and for 81 households in 1969–70 (solid rectangles). Comparison with the histogram on the right demonstrates that, even in the minority of instances when more than one person in a household succumbed to the disease, the cases were mostly contemporaneous not sequential. There was little, if any, transmission under living conditions of the greatest personal intimacy.

We ourselves obtained similar results for the outbreaks of influenza that occurred in English and Welsh boarding schools during the late winter of 1977–78. Although the main feature of our survey was its statistical weight, there were about 10 000 victims in a total school population of about 25 000, in a few schools we were able to obtain precise details of the positions of the beds which had been occupied by victims in the dormitories. The situation for possible transmission was then similar in principle to the households of Hope-Simpson. Like him, we found no evidence that transmission had taken place.

The surprising new feature of our survey was the extreme variability in the incidence of cases according to the physical position occupied by victims within their respective schools. We discussed our findings for two particular schools in Chapter 11. Although pupils in the various school houses met together for instruction, for games, and in some cases for meals, there were nevertheless enormous fluctuations between school houses that were quite beyond what could be attributed to chance.

The facts do not permit any other conclusion than that human activities have little to do with the incidence of influenza. Pandemics sweep over the Earth in their own good time, unhurried by the intensity of human travel or personal contact. The primary agent of the disease evidently lies outside humans, and its attack when examined in detail is exceedingly capricious.

The local details of epidemic attacks on populations distributed over areas with dimensions of a few tens of yards on the low side up to a few hundred miles on the high side accord very well with the meteorology of a storm pattern.

As we have shown elsewhere (*Diseases from Space*, pp. 93–95, Dent, London 1979), the picture is of the primary activating agent being carried downward in water droplets from the tropopause, of the water droplets evaporating before reaching the ground, and of the causative agent of the disease then blowing about the lower atmosphere. To quote from our former discussion:

> 'Strong winds favour the generation of turbulence, although on their own they are by no means sufficient. A wind can blow over a smooth level plane or over the ocean without the generation of turbulence, but a wind blowing over an irregular surface becomes

turbulent if it is strong enough. For wind blowing over the land there are many scales to the turbulence. The largest scale would be induced by major obstacles such as hills and mountains, which would give rise to eddy interchanges up to thousands of feet and would thus be responsible for bringing pathogenic particles (now freed of their raindrops) down to within hundreds of feet of the ground surface. Local irregularities, such as buildings and woodlands, generate intense fine-scale turbulence which can then reach upward for these last remaining hundreds of feet. Local turbulent eddies generated by minor irregularities are, therefore, finally responsible for the detailed incidence of the pathogen at ground level. There may be fortunate spots which are either smooth enough or sheltered enough not to send local eddies upwards through these last remaining hundred feet or so. For these spots, the pathogenic particles will fly by in the wind, immediately above the heads of people who fail to realise their good fortune.'

Reasons why the primary activation agent must be incident from space

The world-wide spread of influenza pandemics cannot be caused by storm patterns, however. Storms are local distributors, whereas the broad global spread of influenza is a larger-scale phenomenon. The distinction corresponds very well to the difference between the lower troposphere of the atmosphere and the higher stratosphere. The tropopause, which separates the lower from the upper regions of the atmosphere, is at a height of about 12 km over middle and high latitudes and at about 17 km over lower latitudes. It is a crucial property of the atmosphere that up to the tropopause the usual situation is for the air temperature to fall with increasing height, to a value of about −60 °C, as shown in Fig. A.2. With increasing height from the tropopause up to about 20 km altitude, the temperature remains close to this value, and then increases steadily to a maximum near 0 °C at an altitude of about 50 km. This region of increasing temperature is the stratosphere, and the altitude of the temperature maximum is the stratopause. Above the stratopause comes the mesophere, which like the lower

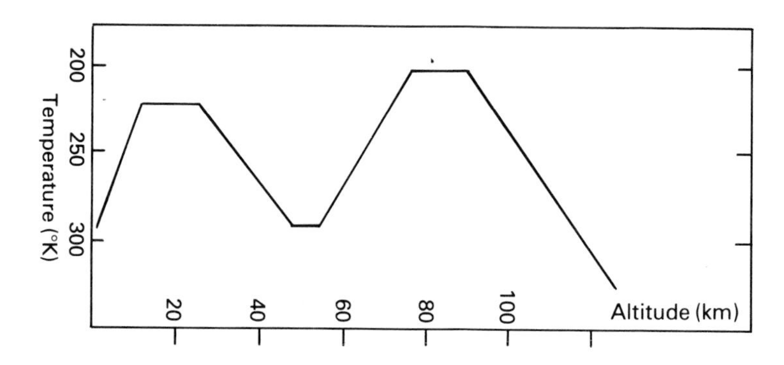

Figure A.2
The variation of temperature with height in the atmosphere.

troposphere is a region in which the temperature decreases with increasing altitude.

The behaviour of the temperature with altitude is relevant to the influenza problem because vertical atmospheric motions of a local kind occur easily when the temperature decreases with altitude (i.e. in the troposphere and mesosphere) but local vertical motions do not occur in the stratosphere. It is only because the atmosphere is heated non-uniformally and because the Earth is spinning that any vertical motions occur at all in the stratosphere, *and such motions as do occur are global not local.*

Consider now the small particles from cometary sources which are entering the Earth's atmosphere at a rate which in aggregate amounts to about 100 tonnes per day. Particles with sizes of 1 μm or less do *not* burn up as they enter the atmosphere. They land 'soft' in the atmosphere at altitudes from 110 km upward. If there are bacteria, viruses and viroids in the cometary material, they can land safely on the Earth, which they could not do on a body without an atmosphere like the Moon.

The atmospheric gases at such high altitudes are very diffuse, and the cometary particles fall under Earth's gravity below the temperature minimum at the 90 km level in a matter of hours. In the mesosphere they are soon carried down to the temperature maximum at the stratopause. This happens because the gases in the mesosphere are constantly in vertical motion.

Further descent from 50 km to 20 km is a different story, however, because the atmospheric density is now much increased so that the time of descent under Earth's gravity is changed from hours to years. Because there is no assistance from local vertical air movements in this region, the stratosphere therefore becomes a trap for the particles.

There are three ways for particles to escape from the stratospheric trap and to come down to the troposphere:

(1) Particles with larger sizes of the order of a micrometre are pulled down to the troposphere in two or three years by Earth's gravity. Viruses or viroids would need either to be clustered or to be encased in a matrix of protective material to take advantage of this mode of descent. In a naked form their smaller sizes would make their times of fall so long, many years, that other modes of descent would be more important.

(2) If a particle contains substances that emit photoelectrons the particle will become positively charged. Quite large electrical fields are generated from time to time in the stratosphere, for example through outbursts of particles from the Sun impinging on the Earth's magnetic field, and through powerful thunderstorms in the troposphere extending their electrical effects into the higher atmosphere. Charged particles, if they are small enough, can be forced downward against the frictional resistance of the atmospheric gases by such fields. In this case it is the smallest particles, as for instance naked small viruses or viroids, for which the effect is strongest. Unlike the other two modes of descent, which are broadly global, involving time-scales of a year or years, this mode is essentially local and instantaneous.

(3) Each year, mostly between latitudes 55° and 65°, there is

> vertical mixing of the stratospheric air. In the northern hemisphere the mixing begins in early November and extends through January. In the southern hemisphere the same phenomenon occurs but with a time displacement of six months. The mixing extends the whole way from the stratopause down to regions immediately above the tropopause (from which a few weeks journey brings the particles down to ground-level).

If we restrict ourselves to particles of the sizes of viruses and viroids, modes (2) and (3) are of main interest. The capricious strikes of mode (2) are in good correspondence with the initiation of the second wave of the 1918 pandemic, while mode (3) corresponds closely to the well-known annual influenza 'season'. As regards mode (2), it is worth noting that outbursts from the Sun are strongly correlated with the 11-year sunspot cycle. In this connection it has been pointed out by Hope-Simpson (*Nature*, 275 (1978), 86) that the pandemics of the twentieth century have tended to break-out at the times of sunspot maxima, which are just the times when solar outbursts tend to be strongest. In *Diseases from Space* (p. 175) we examined the relation of pandemics to sunspots for the eighteenth and nineteenth centuries, finding that Hope-Simpson's association continued to hold good generally, although not quite so precisely as in the present century.

As far as we are aware no sensible suggestion other than mode (3) has been put forward to explain the annual winter influenza season. The alternation six-month apart of winter seasons of the disease in the northern and southern hemisphere is a sure indication of a phenomenon involving the whole of the Earth's atmosphere. Ample evidence of such a six-months displacement is available from influenza records in Australia and S. Africa for the south and from the United States and Europe in the north.

The equator is in a neutral position between the geographical hemispheres. What happens then as the equator is approached from both the north and the south? This question has been partially

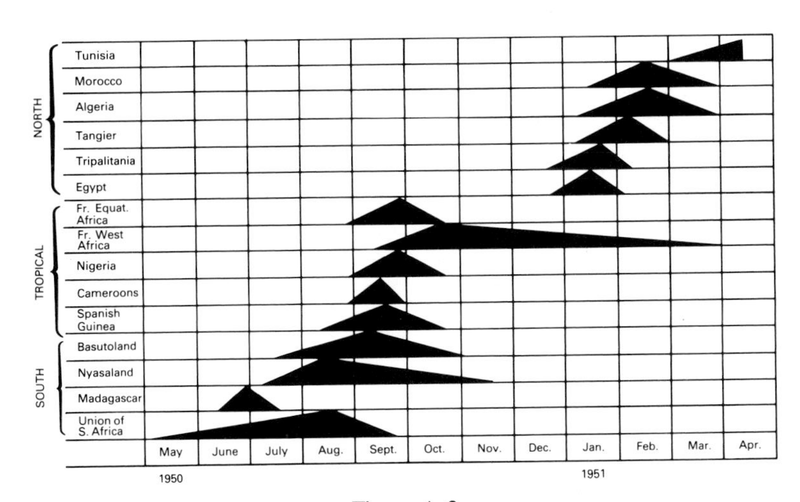

Figure A.3
Hope-Simpson's data illustrating the incidence of an influenza epidemic as a function of geographical latitude.

answered by Hope-Simpson with the aid of data from southern, central and northern Africa for the period May 1950 to April 1951. From Fig. A.3 it is apparent that outside the equatorial band from about 15 °S to 15 °N (contained in the middle section marked 'tropical' in the figure) there is a clear phase lag of about six months between the two hemispheres. Within the 'tropics' thus defined there is a bias towards the south, but with an indication of a transition zone from French West Africa (Senegal, Ivory Coast, Dahomey, at latitudes of about 10 °N). Likely enough, the transition zones are not the same at all longitudes, being biased sometimes south and sometimes north of the equator. Unfortunately, data for latitudes 20 to 25 °N are missing from Fig. A.3, because the lands in these latitudes are desert with low scattered populations. However, the broad features of the transition across the equator can clearly be seen.

Good data for the early winter breakthrough at middle to high latitudes in the stratosphere were obtained from the radioactive isotope rhodium-102 generated in the HARDTACK atmospheric nuclear bomb test of 11 August 1958. The explosion occurred at about 43 km altitude above Johnston Island (16 °N, 170 °W). The nuclear debris from the explosion went overwhelmingly upwards to heights above 100 km, where the radioactive material spread quickly around the whole world. Because the Rh-102 was not particulate, the individual atoms took a time of the order of a year to appear in quantity at the stratopause — unlike even small particles like viroids, the Rh-102 atoms were held up for a

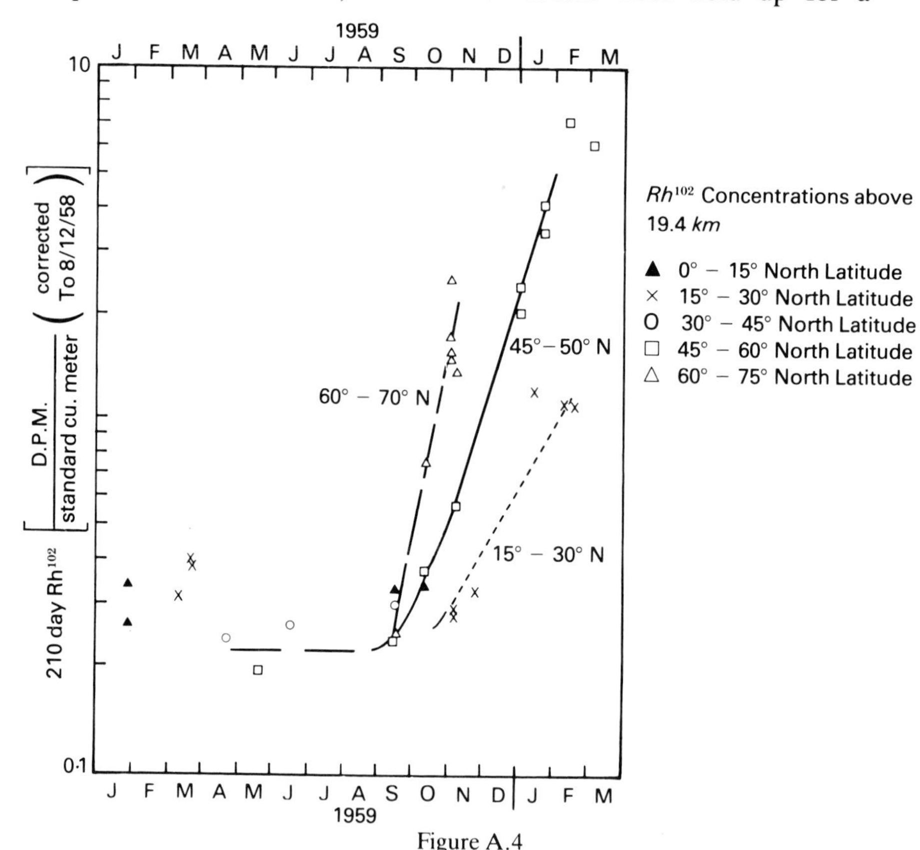

Figure A.4
The fall out of Rh-102 at various latitude intervals from the HARDTACK atmospheric nuclear bomb which was exploded on 11 August 1958.

considerable time in the thin stable air above 90 km. When by mid-1959 they reached the stratopause, however, the atoms of Rh-102 provided a highly effective tracer material for determining global motions downwards through the stratosphere. Data were obtained from samples of air taken at an altitude of about 20 km, which is close to the bottom of the stratospheric trap. From altitude 20 km down to ground-level would be a matter of a further two or three months for individual atoms, but perhaps only two or three weeks for particles. The samples were taken by plane and balloon at various latitudes and longitudes, results being reported by M.I. Kalkstein (*Science*, 137 (1962), 645).

Figure A.4 shows the rising concentrations of Rh-102 that occurred in the northern winter of 1959–60. The primary break down through the stratosphere occurred near latitude 65 °N. This was followed in November to January by a major break at 45 to 50 °N. (The smaller effect at 15–30 °N is not stratospheric, however. It is to be attributed to a horizontal spread of Rh-102 at altitude 20 km taking place from higher to lower latitudes.)

It was possible to estimate the total amount of Rh-102 produced in the nuclear explosion. Comparing the estimated total with the quantities recovered in the samples it was found that it would take of the order of a decade for the whole of the Rh-102 injected into the stratosphere to reach ground-level. This is strikingly similar to the time interval between major influenza pandemics. However, it should be noted that small particles at the stratopause over the extreme polar regions would probably require several decades to clear themselves entirely. The polar caps are in the nature of regions of longer term accumulation.

The epidemiologic data for influenza forces one to take the stratopause as the place of supply to ground-level of the primary causative agent, denoted above by A_1. Because small particles from space arrive inevitably at the stratopause, whereas particles from ground-level would have to fight both Earth's gravity and the general stability of the air from 20 km altitude up to 50 km, it is natural to think of A_1 as being space-incident. The only exception we have been able to think of to this argument would be an ejection of A_1 in the outbursts of volcanos. Although some might prefer a volcanic source to incidence from space, the idea is ruled out by the need of A_1 to have survived in that case for long periods inside the Earth at temperatures in excess of 1000 °C.

Even if we ignore the physical difficulty of A_1 climbing from ground-level up through the stratosphere, a source of A_1 at the ground would long ago have been noticed by epidemiologists.

Specification of a model and the initiation of the disease

We return now to the model,
$A_1 + A_2 + A_3 \rightarrow$ Animal $\rightarrow$ Virus, specifying A_1, A_2, A_3 as follows:

A_1 is a space-incident viroid or genetic fragment capable of 'rescuing' inactivated virus* which has already been added to the respiratory cells of an animal. If A_1 enters a cell and finds no inactivated virus there, it is quickly destroyed by the enzymic apparatus of the cell.

A_2 is a space-incident virus, or portion of a virus, added inactively to the genome of respiratory cells.

A_3 is an inactive virus of terrestrial origin added to the genome of respiratory cells.

It is encouraging that many consequences and insights follow easily from this specification of the model, which requires that influenza can arise only if A_1 enters a cell where either A_2 or A_3 is present already.

Clean Communities

A clean community is one without A_3, without horizontally-

*For an example of a 'rescuing' operation, see M. Park, D.M. Lonsdale, M.C. Timbury, J.H. Subak-Sharpe and J.C.M. Macnab, *Nature,* 285 (1980), 412.

transmitted inactive virus. No person in such a community can contract influenza unless A_1 and A_2 happen to come together in the same respiratory cell. For any one person this would be a highly unlikely event, because with both A_1 and A_2 space-incident and breathed by each person only in small numbers, it is improbable that they will ever come together in the same cell so as to initiate the disease.

Since A_1 is quickly destroyed if it does not find a 'mate', A_1 does not accumulate in the respiratory cells. But the inactivated virus A_2 accumulates, increasing the chance of the disease being contracted. This tendency is limited, however, by the natural death of cells and by their replacement with 'clean' cells. For a solitary individual, never exposed to A_3, such clean replacements of dead cells would keep the probability of contracting the disease very small throughout a whole lifetime. For a community, the probability of a single outbreak of the disease multiples with the size of the population, but if the population is small, limited say to only a few hundred persons, the situation remains that nobody is likely to experience the improbability of A_1 and A_2 coming together in the same cell. Thus a small isolated clean community tends to remain clean.

Dirty Communities

A dirty community is one with A_3. The presence of A_3 implies that someone in the community has experienced the disease. Other members of the community have breathed virus, far larger in quantity than A_2, that has been broadcast by the victim. Because such transmitted virus is hardly ever active, the disease itself was apparently not directly transmitted, in accordance with the epidemiologic facts discussed earlier. Yet those who have been in contact with the victim have now had large numbers of their cells 'primed' by A_3, and will, therefore, be exposed with far higher probability than in a clean community to themselves succumbing, once A_1 comes along again.

The route is evidently opened to an escalating situation. If the first victim spreads sufficient A_3 to infect at least one other person, the virus A_3 spreads eventually throughout the whole community. The community becomes dirty, and likely enough it remains dirty, since once there are many victims the amount of A_3 that is spread around is so large that, short of a decisive immunity developing, the disease becomes self-perpetuating.

All very large communities must be dirty. Even if we suppose an initially clean situation, with each individual having only a tiny probability of A_1 and A_2 coming together to set off the disease, for a large number of individuals the sum of sufficiently many tiny probabilities will add to unity, and an outbreak must happen. Then A_3 will begin to spread progressively, and within months influenza will explode throughout the community, even though there has been no direct case-to-case transmission of the disease.

All young babies are clean, at any rate in the present sense. They are not seriously exposed initially to risk of infection by influenza. Like individuals in a clean community, there is little chance of A_1 and A_2 coming together in their respiratory cells. And because young babies do not usually spend much time in public places, it may be a considerable while, even in a dirty community, before they acquire A_3. The initial cleanliness of somatic cells at birth could be a major factor, both for influenza and for some other diseases, in the seeming immunity of babies. Sooner or later, however, a fond parent or relative googling over the child passes on A_3, apparently without infection taking place. The child is then primed for its first attack of the disease which will inevitably be a bad one because no actual immunity yet exists. This leads to an immense jolt on the immunity system, a jolt which has been described as 'original antigenic sin', a jolt which gives a lifelong bias to the system.

In this connection we note a curious situation that recently came to our notice. During the period 15–20 February 1981 a wave of what seemed like the common cold ripped through a maternity ward at the University Hospital of Wales in Cardiff. Almost all the mothers and nursing staff were affected but remarkably the disease did not seem to pass to the new-born infants, indicating that the new-born infants were not yet primed for attack by the space-incident respiratory virus.

It is inevitable that cometary sources of A_1 must be irregular in their supply to the atmosphere. Because it takes several decades for the whole atmosphere (including the polar caps) to clear itself of any particular injection of A_1, irregularities of supply on time-scales of a few years are smoothed in their incidence at ground-level. If, however, the supply were interrupted on a time-scale of half a century, with the atmosphere being given time to clear itself entirely of A_1, all communities, small and large, would become clean. This would happen through older persons initially with A_3 gradually replacing their respiratory cells, as well as through new births.

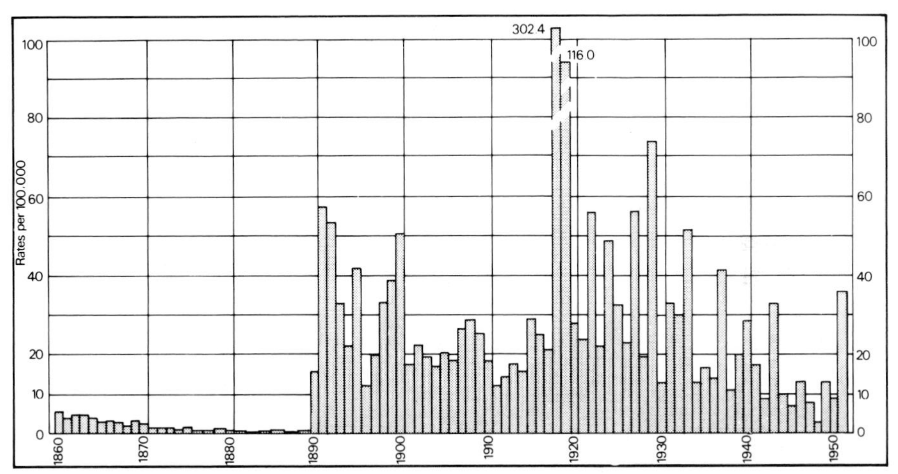

Data Based Partly on the League of Nations Health Organization's 'Annual Epidemiological Reports'

Figure A.5
Annual death rate from influenza in England and Wales

The subsequent resumption of A_1 and A_2, contingent on a new supply from a cometary source into the atmosphere, would then lead, after a slow beginning, to devastatingly explosive outbreaks, especially in densely populated areas, with weakly developed or non-existing immunities causing high mortality rates. Such seems to have been the situation in 1889–91 as shown in Fig. A.5.

A confirmation of the model from a remote geographical source

It was only after we had arrived at the above model that we recalled, from a distant corner of the memory, a paper which had seemed quite inexplicable when we first read it a year or two ago. As one always tends to do with things that seem inexplicable, we had conveniently forgotten about it in the interim.

Before the clearing of the tropical forests of Brazil and Surinam, tribes of Indians lived alone in isolation, typically with populations of only about 500 persons. With the clearing of the forests these tribes were contacted and blood sera were obtained from some of them. From antibody tests, both 'clean' and 'dirty' communities (in the sense we have defined earlier) were found. An example of a 'clean' community was reported by J.V. Neil, F.M. Salzano and P.C. Jacqueira (Studies on the Xavante Indians of the Brazilian Mato Grosso. *Hum. Genet.*, 16 (1964), 52). Similar results were obtained by F.L. Black *et al.* (*Am. J. of Epidemiol,* 100 (1974), 230).

The need for A_2

Although A_1 and A_2 can find each other only very rarely, it would be incorrect to suppose that A_2 plays no role except as a start-up factor in converting an initially clean large community into a dirty one, as may have happened in 1889–91. In cases where A_1 and A_3 set off an attack of the disease, virus particles spread in large numbers throughout the respiratory cells. It then becomes probable that a cell containing A_2 will be invaded by an active particle generated from A_1 and A_3. This may have the effect of 'rescuing' A_2, or of promoting a recombinant event in which a new viable particle is generated partly from A_2 and from the progeny of A_1 and A_3.

If other things were equal from an immunological point of view, such a new particle would be swamped in number by the progeny of A_1 and A_3, and would therefore be of little relevance to the clinical aspects of the disease. But if A_3 persists in a community for a period of years, group immunity will be built up to the point at which few clinical outbreaks occur, although sub-clinical attacks must continue in order that the immunity process be set in operation. Sub-clinical attacks can serve to 'find' A_2, and if either the rescued A_2 or a recombinant virus involving A_2 happens to outflank the immunity system it will be the resulting new viral particle that will persist and grow in number, beyond the stage when the progeny of A_1 and A_3 are brought to a halt in their growth by the immunity system. Thus a new antigenic type of virus will emerge at the clinical level. What happens then is that the new antigenic type replaces A_3 in the community. The disease shifts to a new form which continues until improved immunity to it, together with the arrival of a further A_2, repeats the process.

Work on the H_1N_1, H_2N_2 and H_3N_2 influenza subtypes by C. Scholtissek, W. Rohde, V. von Hoyningen and R. Rott (*Virology*, 87 (1978), 13) strongly suggests that the Asian flu virus H_2N_2 emerged in 1957 as a recombinant generated from H_1N_1 (the previous A_3) and some A_2, while H_3N_2 emerged in 1968 as a recombinant derived from H_2N_2 and some other A_2. Thus A_3 was represented in turn by H_1N_1 (up to 1957), by H_2N_2 (1957 to 1968) and by H_3N_2 (1968 to 1978).

The re-emergence of H_1N_1 in 1977–78 cannot be explained as a recombinant event, however, since the H_1N_1 of 1977–78 was nearly identical with that of 1950.[1] This seems to have been a case in which A_2 (H_1N_1) was rescued substantially without modification, and in which it competed in 1977–78 on about equal terms with the

[1] K. Nakajima, U. Desselberger and P. Palese, *Nature*, 274 (1978), 334.

previous A_3 (H_3N_2). The result was that for a while A_3 had a dual identity. This led 1978–79 to the detection of recombinants between the two forms of A_3 (J.F. Young and P. Palese, *Proc. Natl. Acad. Sci. USA*, 76 (1979), 6547; A.P. Kendal *et al.*, *Am. J. Epidemiol*, 110 (1979), 462).

The need for A_2 to be space-incident

Through the importance attached to the horizontally-transmitted A_3, the present model has come some way towards the usual theory of influenza transmission. The approach towards the usual theory is more apparent than real, however, because A_3, although horizontally transmitted, is considered to be derived from A_1 and from a previous A_2, both of which are space-incident. In effect, therefore, A_3 is also space-incident.

If it could be argued successfully that A_2 had a terrestrial origin, the shift towards the usual theory would be much greater, however. This would accord with the opinion of virologists who have suggested that recombinants are derived in humans from a human component (A_3) and an animal component (A_2), with various preferences given to pigs and birds as the source of the animal component. It has been argued in support of this point of view that segments of human virus can be found that show considerable homology to segments of virus in animals. But if both are built from space-incident components the situation could hardly be otherwise.

If one were to take such a point of view, the main activator A_1 would be space-incident, whereas the source of the virus would be terrestrial, and this would be an uneasy, implausible mixture. Almost inevitably, the next step would be to suppose A_1 also to be of terrestrial origin, but this would lead back immediately to the epidemiologic difficulties set out above. Indeed, one would be returned more or less to the conventional theory with all its attendant problems.

Apart from the epidemiologic difficulties, it would be hard to maintain global genetic uniformity in the influenza virus. The striking aspect of the genetic shifts that occur from decade to decade is that the shifts are everywhere synchronous to within a margin dictated by descent through the stratosphere, synchronous to within a time-scale of months to about a year. If A_2 were contributed by a variety of animals, one might expect a thorough genetic mix-up, with pigs, horses, ducks, chickens, etc., all making their contributions contemporaneously in different geographical locations. Genetic uniformity forces near uniformity in incidence all over the Earth, which points strongly to an extra-terrestrial source for A_2.

Between 1860 and 1889 the death-rate in England and Wales died away almost to zero, and influenza was mild the world over, as seen for instance in Fig. A.5. Were pigs, horses, ducks and chickens everywhere suddenly inactive in their exudents of A_2 over those decades? The question is only a detail, but it is representative of many details which cause trouble in an Earth-bound theory.

INDEX